Innovative Approaches to Undergraduate Mathematics Courses Beyond Calculus

© 2005 by
The Mathematical Association of America (Incorporated)
Library of Congress Control Number 2005930707
ISBN 0-88385-177-6
Printed in the United States of America
Current Printing (last digit):
10 9 8 7 6 5 4 3 2 1

Innovative Approaches to Undergraduate Mathematics Courses Beyond Calculus

Richard J. Maher, Editor
Loyola University Chicago

Published and Distributed by
The Mathematical Association of America

The MAA Notes Series, started in 1982, addresses a broad range of topics and themes of interest to all who are involved with undergraduate mathematics. The volumes in this series are readable, informative, and useful, and help the mathematical community keep up with developments of importance to mathematics.

Council on Publications
Roger Nelsen, *Chair*

Notes Editorial Board
Sr. Barbara E. Reynolds, *Editor*

Paul E. Fishback, *Associate Editor*

Jack Bookman   Annalisa Crannell   Rosalie Dance

William E. Fenton   Mark Parker   Sharon C. Ross

David Sprows

MAA Notes

11. Keys to Improved Instruction by Teaching Assistants and Part-Time Instructors, Committee on Teaching Assistants and Part-Time Instructors, *Bettye Anne Case,* Editor.
13. Reshaping College Mathematics, Committee on the Undergraduate Program in Mathematics, Lynn A. Steen, Editor.
14. Mathematical Writing, by *Donald E. Knuth, Tracy Larrabee, and Paul M. Roberts.*
16. Using Writing to Teach Mathematics, *Andrew Sterrett,* Editor.
17. Priming the Calculus Pump: Innovations and Resources, Committee on Calculus Reform and the First Two Years, a subcomittee of the Committee on the Undergraduate Program in Mathematics, *Thomas W. Tucker,* Editor.
18. Models for Undergraduate Research in Mathematics, *Lester Senechal,* Editor.
19. Visualization in Teaching and Learning Mathematics, Committee on Computers in Mathematics Education, *Steve Cunningham and Walter S. Zimmermann,* Editors.
20. The Laboratory Approach to Teaching Calculus, *L. Carl Leinbach et al.,* Editors.
21. Perspectives on Contemporary Statistics, *David C. Hoaglin and David S. Moore,* Editors.
22. Heeding the Call for Change: Suggestions for Curricular Action, *Lynn A. Steen,* Editor.
24. Symbolic Computation in Undergraduate Mathematics Education, *Zaven A. Karian,* Editor.
25. The Concept of Function: Aspects of Epistemology and Pedagogy, *Guershon Harel and Ed Dubinsky,* Editors.
26. Statistics for the Twenty-First Century, *Florence and Sheldon Gordon,* Editors.
27. Resources for Calculus Collection, Volume 1: Learning by Discovery: A Lab Manual for Calculus, *Anita E. Solow,* Editor.
28. Resources for Calculus Collection, Volume 2: Calculus Problems for a New Century, *Robert Fraga,* Editor.
29. Resources for Calculus Collection, Volume 3: Applications of Calculus, *Philip Straffin,* Editor.
30. Resources for Calculus Collection, Volume 4: Problems for Student Investigation, *Michael B. Jackson and John R. Ramsay,* Editors.
31. Resources for Calculus Collection, Volume 5: Readings for Calculus, *Underwood Dudley,* Editor.
32. Essays in Humanistic Mathematics, *Alvin White,* Editor.
33. Research Issues in Undergraduate Mathematics Learning: Preliminary Analyses and Results, *James J. Kaput and Ed Dubinsky,* Editors.
34. In Eves' Circles, *Joby Milo Anthony,* Editor.
35. You're the Professor, What Next? Ideas and Resources for Preparing College Teachers, The Committee on Preparation for College Teaching, *Bettye Anne Case,* Editor.
36. Preparing for a New Calculus: Conference Proceedings, *Anita E. Solow,* Editor.
37. A Practical Guide to Cooperative Learning in Collegiate Mathematics, *Nancy L. Hagelgans, Barbara E. Reynolds, SDS, Keith Schwingendorf, Draga Vidakovic, Ed Dubinsky, Mazen Shahin, G. Joseph Wimbish, Jr.*
38. Models That Work: Case Studies in Effective Undergraduate Mathematics Programs, *Alan C. Tucker,* Editor.
39. Calculus: The Dynamics of Change, CUPM Subcommittee on Calculus Reform and the First Two Years, *A. Wayne Roberts,* Editor.
40. Vita Mathematica: Historical Research and Integration with Teaching, *Ronald Calinger,* Editor.
41. Geometry Turned On: Dynamic Software in Learning, Teaching, and Research, *James R. King and Doris Schattschneider,* Editors.

42. Resources for Teaching Linear Algebra, *David Carlson, Charles R. Johnson, David C. Lay, A. Duane Porter, Ann E. Watkins, William Watkins,* Editors.
43. Student Assessment in Calculus: A Report of the NSF Working Group on Assessment in Calculus, *Alan Schoenfeld,* Editor.
44. Readings in Cooperative Learning for Undergraduate Mathematics, *Ed Dubinsky, David Mathews, and Barbara E. Reynolds,* Editors.
45. Confronting the Core Curriculum: Considering Change in the Undergraduate Mathematics Major, *John A. Dossey,* Editor.
46. Women in Mathematics: Scaling the Heights, *Deborah Nolan,* Editor.
47. Exemplary Programs in Introductory College Mathematics: Innovative Programs Using Technology, *Susan Lenker,* Editor.
48. Writing in the Teaching and Learning of Mathematics, *John Meier and Thomas Rishel.*
49. Assessment Practices in Undergraduate Mathematics, *Bonnie Gold,* Editor.
50. Revolutions in Differential Equations: Exploring ODEs with Modern Technology, *Michael J. Kallaher,* Editor.
51. Using History to Teach Mathematics: An International Perspective, *Victor J. Katz,* Editor.
52. Teaching Statistics: Resources for Undergraduate Instructors, *Thomas L. Moore,* Editor.
53. Geometry at Work: Papers in Applied Geometry, *Catherine A. Gorini,* Editor.
54. Teaching First: A Guide for New Mathematicians, *Thomas W. Rishel.*
55. Cooperative Learning in Undergraduate Mathematics: Issues That Matter and Strategies That Work, *Elizabeth C. Rogers, Barbara E. Reynolds, Neil A. Davidson, and Anthony D. Thomas,* Editors.
56. Changing Calculus: A Report on Evaluation Efforts and National Impact from 1988 to 1998, *Susan L. Ganter.*
57. Learning to Teach and Teaching to Learn Mathematics: Resources for Professional Development, *Matthew Delong and Dale Winter.*
58. Fractals, Graphics, and Mathematics Education, Benoit Mandelbrot and Michael Frame, Editors.
59. Linear Algebra Gems: Assets for Undergraduate Mathematics, *David Carlson, Charles R. Johnson, David C. Lay, and A. Duane Porter,* Editors.
60. Innovations in Teaching Abstract Algebra, *Allen C. Hibbard and Ellen J. Maycock,* Editors.
61. Changing Core Mathematics, *Chris Arney and Donald Small,* Editors.
62. Achieving Quantitative Literacy: An Urgent Challenge for Higher Education, *Lynn Arthur Steen.*
63. Women Who Love Mathematics: A Sourcebook of Significant Writings, *Miriam Cooney,* Editor.
64. Leading the Mathematical Sciences Department: A Resource for Chairs, *Tina H. Straley, Marcia P. Sward, and Jon W. Scott,* Editors.
65. Innovations in Teaching Statistics, *Joan B. Garfield,* Editor.
66. Mathematics in Service to the Community: Concepts and models for service-learning in the mathematical sciences, *Charles R. Hadlock,* Editor.
67. Innovative Approaches to Undergraduate Mathematics Courses Beyond Calculus, *Richard J. Maher,* Editor.

MAA Service Center
P.O. Box 91112
Washington, DC 20090-1112
1-800-331-1MAA   FAX: 1-301-206-9789

Preface

Initial Comments

This volume has its origins in a contributed papers session, "Innovative Methods in Courses Beyond Calculus," held at Mathfest 2001. This session was organized to discuss the following question:

What can be done to generate and then maintain student interest in the mathematics courses that follow calculus?

Presentations were made by faculty who are addressing this question by doing "something different" in these courses and having a good deal of success with what they are doing. At that time, it was suggested to me that some of the papers presented at that session, when combined with a number of solicited outside papers, might make a useful contribution to the MAA Notes series. The result is this volume, *Innovative Approaches to Undergraduate Mathematics Courses Beyond Calculus,* which contains a wide range of papers that encourage students to take an active role in the learning process and to stretch their learning to ideas and concepts not presented in the classroom.

There is a real need for material of the type contained in this volume, a need that is reinforced by the **CBMS 2000 Survey**. This survey indicates that the number of mathematics majors continues to decline, even though enrollments in Calculus I and II and in the standard second year courses are increasing. Furthermore, a growing number of students in the life and social sciences are seeking more mathematical training in connection with their own disciplines. We should be teaching these students; they should not end up getting this material in courses offered within their own departments with titles like "Mathematical Methods in XX." A similar comment applies to students in the physical sciences and engineering. We have to attract students in all these areas but we are not doing so, even though we are in the midst of an era of increasing college enrollments.

If we are to obtain the results that we desire, we have to rethink what we are doing. We have to make our courses more interesting and more attractive to students. But making courses more interesting is not synonymous with hand waving or watering down content. The authors of the articles in this volume show how we can do all this and still teach **mathematics** courses. These articles introduce new material, offer a variety of approaches to a broad range of courses, and even bring students into contact with ongoing research both in mathematics and in other disciplines. Furthermore, this exposure can take place as early as the second semester of the sophomore year. The articles point out how students can work both individually and in collaboration in order to stretch their mathematical boundaries. This stretching can be accomplished in a variety of ways but the end result is always the same: students leave the course having done something beyond the material covered in the classroom. This volume does have one ongoing theme: it is possible to excite our students about mathematics. But to do so, it is not enough for students to sit and listen to lectures about mathematics. They must actually DO mathematics.

Some Specifics

The multifaceted nature of *Innovative Approaches to Undergraduate Mathematics Courses Beyond Calculus* allows it to address a variety of needs. Faculty trying to revitalize their major offerings will find a host of helpful ideas. Instructors seeking new ways to approach the courses they teach will find a number of models that they can either adopt or adapt. Individuals looking for ideas to incorporate into a specific course will find a wealth of suggestions both in the articles themselves and in their associated references. And teachers who want to expose their students to current mathematical activity will find several avenues to follow. The net result is a volume that offers a variety of options to meet a broad range of needs.

Each of the papers in this volume indicates how the approach discussed can be incorporated into different courses. Specific references are made to twenty-four different upper division mathematics courses, with double references to six of them. These courses include abstract algebra, applied mathematics, biostatistics(two different courses), combinatorics, differential equations, discrete mathematics, game theory, geometry, graph theory, group theory, history of mathematics, linear algebra, mathematical biology, module theory, multivariable calculus, number theory, probability, real analysis, statistics, and topology, along with three capstone courses. In addition, the interdisciplinary applications that are cited involve biology, computer science, economics, engineering, physics, and the social sciences. The reference section of each paper includes additional material. Each article discusses assessment and mentoring. Several of the articles, in addition to the references, also include extensive bibliographies. In general, every effort has been made to make the contents as transparent as possible.

The First Chapter

The first chapter of this volume contains five papers with approaches that are applicable in a variety of courses:

1) *Using Writing and Speaking to Enhance Mathematics Courses*: In this paper, Nadine Myers of Hamline University discusses both writing intensive and oral intensive methods that have been effective in Linear Algebra, Modern Algebra, and Modern Geometry, as well as in structuring a capstone course (Topics in Advanced Mathematics). This approach requires students to write and present proofs frequently and often involves student research on an advanced topic coupled with written and oral presentations. An extensive set of guidelines is included to assist in implementing such an approach.

2) *Enhancing the Curriculum using Readings, Writing, and Creative Projects*: A variety of student-oriented activities can be used to enhance student learning in advanced courses. In this paper, Agnes Rash of St. Joseph's University indicates a number of methods that have been used successfully in Discrete Structures, Group Theory, Number Theory, and Probability and Statistics, as well as in a capstone course involving student research. This article includes a host of specific examples and several extensive reading lists. There is a strong emphasis on student presentations.

3) *How to Develop an ILAP*: The applications of mathematics, while powerful motivators in and of themselves, can have even more impact if they are developed in conjunction with departments that use mathematics within their own disciplines. In this article, Michael Huber and Joseph Myers, both of the United States Military Academy, provide a detailed description of how to work with colleagues in other departments to construct ILAP's(Interdisciplinary Lively Application Projects). The article includes details on how to select a suitable topic, how to work in conjunction with a partner department, what material to include in the student handout, mentoring students during the project, and how to organize student presentations of the results. It also provides several examples of ILAP's and contains references to many others.

4) *The Role of the History of Mathematics in Courses beyond Calculus*: The history of mathematics can provide a good deal of content motivation if it is integrated seamlessly into the course; it cannot appear to be just thrown in. This article, by Herbert Kasube of Bradley University, discusses how such an integration can be accomplished in a variety of courses beyond Calculus, including Abstract Algebra, Combinatorics, Graph Theory, and Number Theory. This paper also provides an extensive bibliography of source material.

5) *A Proofs Course that Addresses Student Transition to Advanced Applied Mathematics Courses*: This paper, by Michael Jones and Arup Mukherjee of Montclair State University, is unique in that it describes a proofs course directed toward a specific curriculum. While emphasizing the construction of proofs, the approach described also encourages students to go through a process that moves from exploration to conjecture to proof in a specific curricular area. It often uses technology to motivate or consolidate ideas.

The Second Chapter

The second chapter of this volume contains five articles that, while more course specific, also contain approaches that are adaptable in other courses.

6) *Wrestling with Finite Groups: Abstract Algebra need not be Passive Sport*: Abstract Algebra is, by its nature, abstract. But it does not have to be approached as a list of definitions and theorems that need to be verified. In this paper, Jason Douma of the University of Sioux Falls discusses how an abstract algebra course can be structured around an open-ended research project. The project is not an application of material covered in class but rather a basis for motivating the actual course content. The paper provides all the details needed to implement such an approach, including information on organization, classroom activity, and assessment.

7) *Making the Epsilons Matter*: Students all too often view an introductory real analysis course as a mechanism that provides the theoretical foundation for calculus results that they already have accepted intuitively. In this paper, Stephen Abbott of Middlebury College describes how an introductory real analysis course can be used to challenge and sharpen intuition as opposed to merely verifying it. He also shows how these outcomes can be reached by a shift of emphasis and not necessarily content, leaving the students with a thorough grounding in the basic concepts of continuity, differentiability, integrability, and convergence.

8) *Innovative Possibilities for Undergraduate Topology*: In this paper, Samuel Smith of St. Joseph's University approaches the undergraduate course in topology as one intended for a broad range of majors and not just those planning on graduate study. To achieve this outcome, the author describes in detail how to structure a course in which an initial geometric approach can be used to motivate the axiomatic structure that characterizes topology. A major goal is to maximize the number of ideas that the students discover for themselves. The possibility of using topology as a capstone course also is explored.

9) *A Project-Based Geometry Course*: Geometry is an axiomatic subject but these axioms need not always be presented in lecture mode. In this paper, Jeff Connor and Barbara Grover, both of Ohio University, discuss a geometry course in which the students develop their own axiom systems, using technology when appropriate. The students receive early experience with both Euclidean and non-Euclidean geometries and also obtain the intellectual tools that they will need to learn any new and unfamiliar mathematics.

10) *Discovering Abstract Algebra: A Constructivist Approach to Module Theory*: Students in upper division mathematics courses can profit from guided discovery, an approach that encourages students to construct their own knowledge and choose their own course of study, while retaining subtle guidance on the part of the teacher to generate definitions, examples, and eventually theorems. In this article, Jill Dietz of St. Olaf's College discusses how to use such an approach in a course on Module Theory that is taught as a follow-up to a first course in Abstract Algebra. It begins with the initial question: "What happens if we replace the vector space axiom that requires an action of a field on an additive abelian group with the

new requirement that there is instead a ring action on an additive abelian group?" The course then uses guided discovery to generate results in what the students eventually find out is module theory.

The Third Chapter

The third chapter of this volume also contains five papers. The first two describe courses relating mathematics and biology, while the third and fourth papers discuss voting theory. The last paper in this chapter discusses a course in Classical Applied Mathematics in which technology is an integral component.

11) *The Importance of Projects in Applied Statistics Courses*: Advanced Statistics courses, particularly those directed toward biology students, no longer are places where formulas are emphasized and data summarized. The focus now is, or should be, on the importance of statistics in providing legitimate answers to the questions posed by researchers. In this paper, Timothy O'Brien of Loyola University Chicago discusses how projects that not only study the assumptions and limitations inherent in research studies but also require student to address statistical topics that are not a part of the standard curriculum greatly expand student learning. These projects may require the analysis of a previously unstudied data set or a critique of original research articles from refereed journals. They also must involve or require the use of techniques that are beyond those covered in the formal classroom presentation.

12) *Mathematical Biology Taught to a Mixed Audience at the Sophomore Level*: Most Mathematical Biology courses are either modeling courses designed for upper division mathematics majors or lower level courses, often with minimal mathematical prerequisites, for biology majors. In this article, Janet Andersen of Hope College describes a team taught course in Mathematical Biology that serves both mathematics and biology majors. The prerequisite for the mathematics majors is a course in Linear Algebra and Differential Equations while the biology majors are required to have one semester in Calculus plus a sophomore level course on Ecology and Evolutionary Biology. The course is based on biology research papers that use matrix analysis or differential equations in their development. Class requirements include both collaborative work and classroom presentations.

13) *A Geometric Approach to Voting Theory for Mathematics Majors*: Voting theory can be incorporated into a variety of upper division mathematics courses, which can allow students to obtain some insight into ongoing mathematical research and its outcomes. In this article, Tommy Ratliff of Wheaton College discusses how this can be done in a course that also covers game theory. With discrete mathematics as a prerequisite, this course delves into the geometric framework that underlies some of the recent results obtained in voting theory. One outcome is to help students become better judges of the choice procedures available to them in their everyday lives.

14) *Integrating Combinatorics, Geometry, and Probability through the Shapley-Shubik Power Index*: Voting theory is a rapidly developing area of mathematics with a broad range of applications both inside and outside of mathematics. This paper, by Matthew Haines of Augsburg College and Michael Jones of Montclair State University, serves as a primer for instructors who wish to introduce the elements of voting theory into their courses. This article also discusses how its contents can be applied in several upper division courses.

15) *An Innovative Approach to Post-Calculus Classical Applied Math*: Classical Applied Mathematics typically is considered to be a formal development of the theorems and problem-solving techniques of applied analysis. In this article, Robert Lopez, formerly of Rose-Hulman Institute of Technology and now with Maplesoft, indicates why a computer algebra system should be the working tool for teaching, learning, and doing classical applied mathematics. The result of such an approach is a richer, more efficient, and more effective learning system. One key point is that technology must be an integral part of the course that is available for all parts of the course, including examinations. Two in-depth examples are provided to illustrate the effectiveness of this approach.

Conclusion

Innovative Approaches to Undergraduate Mathematics Courses Beyond Calculus is intended to serve as a starting point both for those who plan to adopt or adapt the approaches it discusses and for those who plan to develop their own ideas. This volume contains fifteen papers that provide useful information on alternative methods that are being used with great success in the courses following calculus. These papers not only present new concepts and related applications that can be introduced into these courses but in several cases also bring students into contact with ongoing research. In all cases, the material is presented in detail. When the presentation is course specific, techniques for using the approach in other courses also are discussed. Much of the heavy lifting already has been done and all that remains is for instructors to adapt the suggestions to their own individual settings.

Hopefully, this volume will provide a body of information that will prove helpful to instructors teaching the courses that follow calculus. At this time, there does not appear to be anything in print that discusses how to generate and maintain student interest in the courses beyond calculus. While *Innovative Approaches to Undergraduate Mathematics Courses Beyond Calculus* may well be the first of its kind, it most certainly will not be the last. The need to make our major programs more attractive and to draw students outside our discipline into our courses will see to that.

Contents

Preface vii

Chapter 1 Papers Covering Several Courses

Introduction 3

1.1 Using Writing and Speaking to Enhance Mathematics Courses, *Nadine Myers* 5

1.2 Enhancing the Curriculum Using Reading, Writing, and Creative Projects, *Agnes Rash* 13

1.3 How to Develop an ILAP, *Michael Huber and Joseph Myers* 23

1.4 The Role of the History of Mathematics in Courses Beyond Calculus, *Herbert Kasube* 33

1.5 A Proofs Course That Addresses Student Transition to
    Advanced Applied Mathematics Courses, *Michael Jones and Arup Mukherjee* 39

Chapter 2 Course-Specific Papers

Introduction 55

2.1 Wrestling with Finite Groups; Abstract Algebra
    need not be a Passive Sport, *Jason Douma* 57

2.2 Making the Epsilons Matter, *Stephen Abbott* 67

2.3 Innovative Possibilities for Undergraduate Topology, *Samuel Smith* 81

2.4 A Project Based Geometry Course, *Jeff Connor and Barbara Grove* 89

2.5 Discovering Abstract Algebra: A Constructivist Approach to Module Theory, *Jill Dietz* 101

Chapter 3 Papers on Special Topics

Introduction 113

3.1 The Importance of Projects in Applied Statistics Courses, *Timothy O'Brien* 115

3.2 Mathematical Biology Taught to a Mixed
    Audience at the Sophomore Level, *Janet Andersen* 127

3.3 A Geometric Approach to Voting Theory for Mathematics Majors, *Tommy Ratliff* 133

3.4 Integrating Combinatorics, Geometry, and Probability
    Through the Shapley-Shubik Power Index, *Matthew Haines and Michael Jones* 143

3.5 An Innovative Approach to Post-Calculus Classical Applied Math, *Robert Lopez* 163

About the Editor 173

Chapter 1

Papers Covering Several Courses

Introduction 3

1.1 Using Writing and Speaking to Enhance Mathematics Courses, *Nadine Myers* 5

1.2 Enhancing the Curriculum Using Reading, Writing, and Creative Projects, *Agnes Rash* 13

1.3 How to Develop an ILAP, *Michael Huber and Joseph Myers* 23

1.4 The Role of the History of Mathematics in Courses Beyond Calculus, *Herbert Kasube* 33

1.5 A Proofs Course That Addresses Student Transition to
Advanced Applied Mathematics Courses, *Michael Jones and Arup Mukherjee* 39

Introduction

The first chapter contains five papers with ideas, approaches, and applications that range over several different areas. The articles by Nadine Myers of Hamline University and Agnes Rash of St. Joseph's University show how oral presentations, projects, readings, and writing can be used creatively to enhance student learning and interest in courses like Discrete Mathematics, Linear Algebra, Modern Algebra, Number Theory, Probability, and Statistics. Both articles emphasize student cooperation and participation and both contain extensive information on implementation and mentoring.

Mathematics is used extensively in other disciplines and the article by Michael Huber and Joseph Myers of the United States Military Academy describes how to take advantage of this largely untapped resource. They describe in detail how to develop ILAP's (Interdisciplinary Lively Application Projects) by working in conjunction with faculty in science, engineering, and the social sciences. Once a project is prepared, it is assigned to students for group work; some projects are usable as early as the first semester of the sophomore year. While several ILAP's are described and the bibliography contains extensive references to others, the major goal of the article is to provide extensive information on how to design and implement an ILAP.

Many upper division mathematics courses spend little time indicating how the history of mathematics has influenced the course content. In his paper, Herbert Kasube of Bradley University notes how an understanding of the history of a subject, when integrated into the course, can motivate students to pursue the subject matter. In addition to indicating how this can be accomplished in several different upper division courses, this article also contains an extensive list of source material.

Transition courses that introduce students to proofs were quite popular in the past. They fell out of favor for a while when departments began to use introductory courses in number theory, geometry, or, especially, linear algebra, in their place. Transition courses are now making a comeback as departments once again discover that students who have completed a calculus sequence often are not prepared for what they encounter in introductory algebra or analysis courses. The article by Michael Jones and Arup Mukherjee, both of Montclair State University, surveys the various directions that such courses can take. It then focuses on a unique approach to a transition course, one that requires students to use exploration and technology to conjecture and then prove a variety of results in applied mathematics.

The five articles in Chapter 1 describe approaches that can be used in a variety of settings. They are unified around the theme of encouraging students to look beyond the course material to find something more, a theme that remains constant throughout this volume. They use everything from the history of mathematics to modern technology to motivate student learning and insight. The goal of these articles, along with those in Chapters 2 and 3, is to generate student interest and appreciation of mathematics and its applications.

1.1

Using Writing and Speaking to Enhance Mathematics Courses

Nadine C. Myers
Hamline University

1.1.1 Introduction

A college wide curriculum change in 1986 prompted my adoption of nontraditional methods in mathematics courses. The curriculum requires students to take one writing intensive course per year and two oral intensive courses in four years. Good practice suggests that students enroll in at least one class of each type in their major field. Given this need for writing intensive or oral intensive mathematics courses, I first developed a writing intensive and oral intensive Modern Geometry course [11]. Later I developed an oral intensive Modern Algebra course [12] and have used similar techniques in Linear Algebra and a capstone course, Topics in Advanced Mathematics. In all these courses, I have two main goals: 1) Make the class time as discussion oriented as possible, and 2) Require students to write and present proofs frequently. In Modern Geometry and in Topics, students also must research an appropriate topic and then write and present a paper on it. Using these methods, I observed an increase in both student satisfaction and success. Following a brief description of writing and oral intensive courses, this paper will describe my methods.

A writing intensive course must provide students with instruction on appropriate writing in the discipline and require them to complete multiple writing assignments. Faculty must provide regular feedback and allow students to reflect on and revise their writing. A final requirement, which has in my experience proved to be a consequence of the others, is that significant learning must occur as a result of students' writing activities. Analogous requirements pertain to the oral intensive designation. I interpret writing in the discipline for mathematics to mean both exposition and proof writing, including appropriate use of symbols and technical terms. Oral intensive mathematics courses also require multiple opportunities for presentation, regular feedback, and instruction in oral skills such as choosing cogent, precise language, and pronouncing symbols correctly.

1.1.2 Teaching Students to Write Mathematically

To facilitate the writing intensive aspect of Modern Geometry, I introduce Reader Expectation Theory (RET) early in the term. Briefly, this theory helps students to write in such a way that individual components of a given document are located where the reader expects to find them, thereby minimizing structural confusion and maximizing substantive clarity. RET encourages writers to use the active voice. Students are taught to ask: What is the action in this sentence? Is that action expressed directly and clearly as a verb? Who or what is the agent of that action? Is that agent *who* or *what* the sentence is really about? What is my point in this sentence? Is it immediately before the period, i.e., in the *stress* position? How does this sentence connect with the previous sentence and with the next? Is this sentence in the appropriate place relative to its importance to the topic of the paragraph? How does this paragraph fit into the structure of the whole document? (See [7], and [15] for information on RET and clarity.)

RET gives students a clear format for editing their papers. It makes their writing more intentional by requiring them to think of what they want the reader to take away from each sentence, from each paragraph, and from the entire document. Beyond helping students write clear expository papers, I believe that RET helps students write proofs. To write an effective proof, a student should ask: What is happening in this line of the proof? What is the agent or cause of that action? Has that cause been sufficiently exposed? What is my purpose in this line? What is my point in the entire proof? Once I am clear on the point sentence, and have located it as the final line of the proof (stress position), how shall I organize the steps between the hypotheses and the conclusion?

Teaching Students to Write Proofs

In my courses I devote some time to teaching students how to find and write proofs. Of course some students learn by imitating the professor, and most of them have taken a transition course in sets, logic and proof techniques. But even the best students benefit by explicit attention to proof writing, which proceeds generally as follows. First I ask students to identify the if-then form of the proposition. Once they are clear on the hypothesis and conclusion, I have them write the conclusion at the bottom of the page as the final statement of the proof. Then they write the hypotheses at the top of the page. Also at the top they write an interpretation of the hypotheses: What do they mean? What are some consequences or implications of the hypotheses? At the bottom they write an interpretation of the conclusion: What does it mean? (Since students only gradually recognize the value of precise definitions, I want them to have every relevant definition clearly written on one end of the paper or the other.) Then I ask: What are some ways of arriving at the conclusion, given the hypotheses? Are there any relevant theorems? Then they try to make connections. Once they perceive the main line of argument, they must fill in the middle of the page one step at a time, justifying each step by one of the six classical reasons: By hypothesis, by definition, by axiom, by theorem, by rule of logic, by a previous step. They may also use "by properties of numbers" to justify algebraic or numerical manipulation. Although students initially complain about the need for justification, they soon recognize that the discipline of supplying reasons helps to clarify their thinking.

Writing and Speaking Activities

In Modern Algebra, Topics, and Modern Geometry I give multiple proof writing and presentation assignments. The three classes are enough alike that I will limit discussion in this paper to the Modern Algebra course. Similar methods are also used, though less extensively, in Linear Algebra. I use collaborative proof writing and presentation partly as a means of implementing oral intensive requirements but mostly because they are an effective means for teaching the material. Students are asked to work in groups for both writing and presenting proofs. They may also work together on homework assignments. Most groups recognize that they produce work that is significantly better than any individual could do. That students regard group problem solving as beneficial is noted in the student evaluation comments listed below. That groups provide an effective way to learn is discussed in [5], [6], [11], [14], and elsewhere.

In Modern Algebra, eight class days are reserved as presentation days. Class time during the remaining days is used for informal lecture, questions, and group discussion. As the term progresses, lecture time diminishes as questions and group discussion time grow. Group discussion can take many forms. Sometimes we analyze proofs in the text. I guide students through the author's proof, asking them to identify how the hypotheses are used, asking them to supply reasons for steps that the author has not explicitly justified, or asking them to identify the conclusion and its connection to what the author actually stated in the final line of the proof. In the latter case, usually for a proof by contraposition, I find that some students can verify each step of the proof yet remain unable to explain why or how those steps prove the theorem. Other group discussions are for answering student questions. If one student is stuck on a problem, others are asked to suggest a way to resolve the immediate difficulty or to suggest an alternative approach. Finally, discussion time may be used for solving problems posed by the instructor. I might ask why a given theorem's conclusion is what it is, and ask students what prevents the conclusion from being stronger. For instance, why is a given factor ring only a commutative ring with identity? What prevents it from being an integral domain? How might the hypotheses be changed to bring about the stronger conclusion?

I regard group problem solving outside of class as a crucial speaking activity. As we proceed through the textbook, I assign problems daily. Some are for presentation, with the remainder to be written up and handed in. Almost all require proofs. Students work with their group to solve the problems and may consult with the instructor as needed; they even can submit trial proofs for feedback. For final submission, every student must write up every problem. Occasionally, problems are graded on a group basis. That is, I grade problem 1 on one student's paper (chosen randomly), problem 2 on the next, and so forth, advancing group members' papers cyclically after each problem. Thus a given problem is graded only once per group, with all members receiving the same score.

Assigned problems that are not submitted for grading are used on presentation days. At the beginning of the class period on such days, each group lists the problems it is able to do. A problem cannot be listed unless every group member is capable of presenting its solution. Groups are expected to master at least eighty per cent of the assigned problems. Problems range from easy to quite difficult in my judgment, the majority being about medium difficulty. Each group presents to the class at least one problem from their list. Early in the term students in non-presenting groups are asked to evaluate each proof using the form in Appendix A. The form relates to the proof writing technique discussed above. Non-presenting students are graded on their ability to recognize that the presenter has or has not followed the suggested form. As the term progresses, the evaluation form is abandoned in favor of proof critiques by non-presenters.

To accommodate students' growing capabilities during the term, presentation requirements also grow. Early on, I invite each group to select its best proof for presentation, subject only to avoiding overlap among the groups. One member of each group volunteers to write the solution on the board, coached by others in the group. To save time, there are multiple groups at the blackboard, writing solutions simultaneously. (Or multiple groups writing their proofs on transparencies for the overhead projector.) Once the solutions are all written, a different representative explains her or his group's solution to the class. All members of a presenting group are encouraged to monitor the written solution and explanation to ensure accurate transmission of the group's work.

Once a solution has been explained to the presenting group's satisfaction, the remaining class members are asked for their evaluation: Is the proof valid? Are the hypotheses and conclusion stated and interpreted correctly? Is there a logical progression from hypotheses to conclusion? Has the group used correct logic, terminology, and symbols? Is the proof concise? Could it be better organized? Is there an alternate approach? Why might someone choose one approach over another? As the term advances and students become more comfortable with proofs and presentations, selection of problems and presenters becomes more focused. By the third or fourth presentation day, I select a problem for each group and an individual to present the problem. Group members may still coach the presenter as she or he writes the solution on the board and may respond to audience questions if the designated presenter is stymied. Toward the end of the term, group coaching is expected to diminish, although group members are still given the first opportunity to correct any mistakes in a proof. For the final presentation day, students solve problems with their groups, but each individual submits the three most difficult (in their judgment) problems that she or he is willing to present. The instructor assigns one problem per student for individual presentations. If there is insufficient time for all students to make a final presentation to the class, the remaining presenters schedule individual appointments with the instructor.

1.1.3 Grading

Since Modern Algebra is an oral intensive course, students are graded on their participation in class discussion. The general grade guidelines given in Appendix B are used. During class discussion, one must prevent the more aggressive students from dominating and encourage timid students to contribute. In my experience this has not been a real problem. If it were, I would follow a physics colleague's advice: Simply let the class know that everyone can get an A in discussion for the day, but no one gets an A unless everyone's voice has been heard in a meaningful way.

In the algebra course, final grades are based on performance in four areas: Presentations, participation, problem sets, and examinations. Relative weights are:

Presentations 150
Class participation 100
Problem sets 200
Hour examinations (2) 200
Final Examination 150

On presentation days, every student gets two grades: a presentation grade, and an observer grade. Presentation grades are group grades, based jointly on the number of problems that the group can do, together with a grade on the quality of the presentation. A presentation is graded A, B, or C according to the proof's correctness and quality. I have never seen a presentation deserving a D or F. Groups work hard to get all members up to speed on the problems. The grades are then scaled according to the proportion of assigned problems done by the group. For example, if a group scored an A on a given problem presentation, but could do only sixty per cent of the assigned problems, their presentation grade for that day would be B. Forty per cent or fewer would result in a C or D. Every member of the group gets the same presentation grade. An observer for a particular presentation is a class member who is not part of the presenting group. Observers critique a proof as described above and are graded according to the guidelines in Appendix B.

1.1.4 Conclusions

Below are some reactions by students and myself.

Instructor's Observations

The Modern Algebra course described here took some time to develop and is described in [12]. The course as it now exists works very well. In twenty years of teaching, I have not seen students capable of handling algebraic structures like factor rings and field extensions with such confidence nor proofs written with such clarity. Although there may be any number of reasons for this occurrence, some are clearly related to the oral intensive format. Some reasons for success are the usual benefits of group work: shared insights, opportunities to ask questions as they arise during problem solving, peer review of preliminary work, and motivation to persevere for the sake of the group. Judging from the camaraderie and good-natured banter that occurred in class, especially on presentation days, students enjoy as well as benefit from group problem solving and presentation.

Another reason for success is that eight presentation days along with weekly written problem sets provide students with a constant incentive to work. Students need to discuss algebra frequently within their groups in order to keep up. As the course evolved it became clear that proof presentations and the associated critiques contribute significantly to students' success. As indicated by their course evaluations, students themselves recognize that presenting work to the class helps them learn. If a student produces an invalid argument, suggestions from other students almost always leads to a solid proof. While students are initially reluctant to criticize other students' work, encouragement to make a proof even better seemed to draw out a few more comments on each successive presentation. Also, I make a point of cautioning students that failure to identify flaws in another student's proof will be understood as their belief in its validity.

Students know that their participation score will be affected by their ability to find flaws as well as their willingness to offer clearer language, better notation, or alternate methods. Proof critiques improve dramatically as the term progresses, and there are many excellent discussions of proofs. As students offer suggestions to other students, it quickly becomes clear to everyone that they all can write good proofs. In one class, two students who had initially dubbed themselves hopeless at proving things found that they were able to construct a fair argument by the end of the term.

Student Reactions

Most students respond well to the course. Below are a few comments taken from anonymous student course evaluations. All are responses to the question, "What features of this course most effectively helped you learn?"

> If I presented a problem I really knew how to do it by the time I was done.
> Math is usually pretty lecture intensive, but I appreciated more active learning through the presentations.
> I ... also enjoy the group collaboration on problem sets.
> Presentations [helped me learn] — good way to clarify the thought process.
> Working in a group was most beneficial for me.
> The group sessions helped me learn and I depended on this to help me understand.
> A number of students expressed specific satisfaction with the course.
> I have enjoyed [this course].
> We are encouraged to ask questions and to participate in the learning process.
> Very interesting.
> Stirs such enthusiasm.
> Very well planned and taught.

The few negative comments centered on the necessity for students to attend class regularly (a requirement for their grades for class discussion). A few students commented negatively on the workload. One student said that the large number of assigned problems helped him learn, but that "more problems should be voluntary, thus making the student more responsible for wanting to learn more by doing more."

Appendix A

Author's note: Students are initially encouraged to write proofs by carefully considering four questions: What are the hypotheses? What do the hypotheses mean? What is the conclusion? What does the conclusion mean? Then they are asked to supply logical connections between the hypotheses and conclusion. Proof presentations are scored by students according to the presence or absence of these items.

Evaluation Form for Presentations

Evaluator's Name:

Use the following scoring for the first four items:
- 0 = item missing
- 1 = item partially present
- 2 = item complete

For the Connection item value the item as appropriate between 0 and 4:
- 0 = proof makes no connection between hypothesis and conclusion
- 4 = complete logical connection made

Presenter or group name	Hypothesis stated	Hypothesis explained	Conclusion stated	Conclusion explained	Connection logically made
1.					
2.					
etc.					

Appendix B

Author's note: A copy of the following statement is distributed at the beginning of the term.

Grading Class Participation

Class Participation is crucial to success in this class. Participation means showing up for class having completed the assigned reading, asking questions about anything in the reading or assignment that seems unclear, offering insights, and listening to the comments, questions or insights of others. It also involves proof critiques on presentation days.

Evaluation of participation falls into the following categories:

A-range
    Regularly makes helpful, relevant contributions to discussion.
    Regularly asks questions clarifying questions about text material or problems.
    Frequently offers suggestions or insights that advance an argument or allow others to understand the material better.

B-range
    Occasionally makes helpful, relevant contributions to discussion.
    Asks occasional questions.
    Occasionally offers suggestions or insights that advance an argument or allow others to understand the material better.

C-range
> Attends class regularly and actively pays attention to discussion.
> Occasionally contributes ideas, answers, or suggestions for improvement.

D or F-range
> Does not attend regularly.
> Does not pay attention to discussion.
> Seldom contributes to discussion.

Modifiers
> Missing more than three classes will lower your grade.
> Being totally distracted or inattentive will lower your grade.

Making contributions to discussion means:
> Asking questions about things in the text, or things said in class, which are unclear or confusing.
> Offering answers to questions asked by others in class.
> Making claims or observations about the ideas being discussed.
> Offering alternative arguments or ideas.
> Offering support, criticism, modification, or clarification for proofs being discussed. This includes the appropriate use of mathematical language and symbols.
> Asking questions about or pointing out possible flaws in presented proofs.

Notice that the sheer number of your contributions does nothing to improve your grade. Contributions must be 1) relevant and 2) helpful. A genuine question always counts as relevant and helpful.

Relevant contributions show that you are engaging with the concepts being discussed at the time, and that you are well prepared for class.

> Helpful contributions advance or improve the discussion by bringing in new ideas or insights.
> Helpful contributions also let us understand the ideas or arguments being put forward.
> Relevant contributions point out alternate approaches to the problem or concept being discussed.

References

1. T. Barr, "Integrating Mathematical Ideas Through Reading, Writing, and Speaking: A Senior seminar in Mathematics and Computer Science," *PRIMUS* 5 (1995) 43{-54.
2. R. Burn, "Participating in the Learning of Group Theory," *PRIMUS* 8 (1998) 305–316.
3. S. Buyske, "Student Communication in the Advanced Mathematics Classroom," *PRIMUS* 5 (1995) 23–32.
4. R. Czerwinski, "A Writing Assignment in Abstract Algebra," *PRIMUS* 4 (1994) 117–124.
5. N. Davidson, *Cooperative Learning in Mathematics: A Handbook for Teachers*, Addison Wesley, New York, 1990.
6. J. Gersting and J. Kuczkowski, "Why and How to Use Small Groups in the Mathematics Classroom," *CMJ* 8 (1977) 270–274.
7. G. Gopen and D. Smith, "What's an Assignment Like You Doing in a Course Like This?" in *Writing to Learn Mathematics and the Sciences*, Teachers College Press, New York, 1989.
8. N. Hagelgans, "Active Learning in Abstract Algebra," presentation at Joint Mathematics Meetings, 1999.
9. W. Martin, D. Vidakovic, and R. Vakil, "Proving it in small groups: Collaborative Searches for Mathematical Proofs," presentation at Joint Mathematics Meetings, 1999.
10. R. Maher, "Step By Step Proofs and Small Groups in First Courses in Algebra and Analysis," *PRIMUS* 4 (1994) 235–243.
11. N. Myers, "Writing and Speaking to Learn Geometry," *PRIMUS* 1 (1991) 287–294.
12. ——, "An Oral Intensive Abstract Algebra Course." *PRIMUS* 10 (2000) 193–205.

13. National Research Council, *Moving Beyond Myths: Revitalizing Undergraduate Mathematics*, National Academy Press, Washington, 1991.
14. N. Hagelgans, B. Reynolds, K. Schwingendorf, D. Vidakovic, E. Dubinsky, M. Shahin, G. Wimbish, Jr., *A Practical Guide to Cooperative Learning in Collegiate Mathematics*, The Mathematical Association of America, Washington, 1995.
15. J. Williams, *Style: Ten Lessons in Clarity and Grace (2nd Ed)*, Scott Foresman Company, Chicago, 1985.

Brief Biographical Sketch

Nadine Myers received her PhD from the University of Iowa. Currently she is a professor and chair of the mathematics department at Hamline University in St. Paul, Minnesota. Besides pedagogy, her mathematical interests include abstract algebra and graph theory.

1.2

Enhancing the Curriculum Using Reading, Writing, and Creative Projects

Agnes Rash
St. Joseph's University

1.2.1 Introduction

Theoretical upper division mathematics courses may be interesting in and of themselves, but making them lively can be challenging. This article discusses the use of readings, student projects, and other creative endeavors to enhance understanding and make these courses come alive. A large part of the success of these extensions is providing the opportunity for students to discuss and explain what they have learned or accomplished during the course. Students react enthusiastically to the readings and projects. Upper division courses can become a natural place to introduce modern applications of mathematics. Specific examples of student research presented here are primarily drawn from number theory, discrete structures, group theory, and probability and statistics. Samples of suggested reading lists are provided. This article provides two examples of how the suggestions come together to form a cohesive course in the sections entitled **Putting It All Together**. Finally, the culmination of a project deserves special attention. At the end of the article is a section on **Showcasing Student Accomplishments** that presents rewarding opportunities for students to distinguish themselves.

As mathematics departments become more involved with capstone courses for their students, the information in this paper may be helpful in designing a capstone experience for your students. In this seminar each student, under the guidance of a faculty mentor, undertakes an independent project culminating in a presentation in the department. The topic may be suggested by the student or chosen by the mentor. Often the topic is an extension of material covered in an upper division mathematics course or is an interdisciplinary application of mathematical content. The venue for presentation depends on departmental preferences and varies from one institution to another. Some suggestions in this direction are given in the section on **Showcasing Student Accomplishments**.

1.2.2 Reading about the Subject

For any upper division course, consider developing a reading list. For many students, their experiences in reading mathematics have been limited to reading the textbook or, even worse, reading just the definitions, theorems, and a few examples in the text. Encourage students to select readings of interest and to submit written critiques of the articles. These articles can be an extension of a topic discussed in the course, an application of a result to another field, or an historical note about the development of the concept. Students can select readings that are appropriate to their own level of development or the instructor may make particular suggestions about topics for individuals. While a general reading list is helpful, a specific reading list for each course is more beneficial. Five journals that regularly carry articles suitable for undergraduate students to read are: *Math Horizons, Pi Mu Epsilon Journal, Chance, Stats* and *The Journal of Recreational Mathematics*. Construct a reading list that contains a range of articles to suit the varied interests of students. A reading list for a course is a work in progress. Each year, new and interesting articles appear, while older articles may need to be deleted from the list. For example, before Andrew Wiles proved Fermat's conjecture there were many articles referring to the following theorem as unproven: $x^n + y^n = z^n$ has no solution for n greater than 2. And vary the level of difficulty from easy (to be read by everyone) to demanding. Stretching the bright students beyond the content of the course can challenge them to learn concepts on their own.

After a student has read an article, s/he should submit a report to the instructor. This report can be a synopsis of the article, a critique of the article, or any written document that the instructor believes would provide an accurate representation of the student's comprehension of and interest in the material. These critiques often are helpful to the instructor. The written content can be an indicator of the comprehension of the ideas presented. Sometimes the instructor can ascertain a student's interests by reading his or her summary of the reading. The instructor may then use this information to guide that student into a suitable topic for research.

1.2.3 Student Projects

Projects can serve several purposes. They offer the instructor the opportunity to broaden the range of theory or applications . Projects can provide students with challenging problems involving concepts beyond the scope of the lectures. Designing and solving problems creates an environment similar to the one that they may encounter in a real world setting after graduating from college. Real world problems involve first reading about the area, deciphering the information into a manageable form, finding a solution to the problem, and then explaining the findings in a written or oral presentation. The use of projects, either individual or group, improves the ability of students to communicate mathematical concepts to their peers. Students differ in their interests and abilities, so finding suitable problems can be challenging.

When determining suitable topics for students, it usually is desirable to interview the students to determine their interests. Discuss possible projects that match these interests. Here is an opportunity to be inventive. Frequently students do not see that a mathematical model may be appropriate to a particular situation. Music or sports are common areas of student interest and both areas lend themselves to mathematical models. For example, consider the process, or art, of juggling, which involves keeping multiple objects in the air at the same time. Various juggling patterns can be modeled using linear algebra (eigenvalues and Markov chains) and graph theory. Students will be able to find literature on the topic. You may suggest some relevant readings to help students focus on a problem.

Give each student the opportunity to think about your conversation and to read about the problem before scheduling another meeting. If you cannot determine a relationship between the student's interest and a suitable mathematical problem, consult your colleagues. Many creative ideas may emerge that you can share with your student. Try to foster creative thinking in each student. There are no bad ideas when students are trying to find a topic. Even though some topics are more difficult and perhaps beyond the scope of the student's knowledge or ability, the student's interests still are valid.

During a second interview, the faculty member and the student can decide on the project to be completed. The process of mentoring a student involves a commitment of both parties to frequent discussions about progress on the work. Students need encouragement and occasionally a little nudge to stay focused. The mentor's task is to encourage the student, perhaps critique the written work, but not to solve the problem or write the results. Constructive suggestions often are necessary and can be provided in a positive manner. Be patient, but always keep the deadline in view. The problem may have to be restricted (via simplifying assumptions or restricting of the scope of the problem) in order for the student to complete the project in a reasonable amount of time. Some excellent project ideas were suggested during the session "Mathematical Modeling In and Out of the Classroom" at the AMS/MAA joint meeting in January 2003 [4].

In a course on problem solving or topics in mathematics, students can study elementary Euler circuits and Hamilton circuits. Finding the shortest path from one campus building to another is a possible group undertaking. The shortest path can be defined in terms of distance traveled or shortest time needed to traverse the route. An explanation of continuity and its importance in the real world, the problems presented by discontinuities, is a project that requires only a background in calculus. Projects from abstract algebra include computing all groups of a specific order n. Students also may explore why $A_5, A_6, \ldots$ are simple groups and why mathematicians are interested in them. Depending on the emphasis of the course, many other opportunities exist to explore concepts. In group theory, for example, students can also explain how to construct the semidirect product of two finite groups, G and H, using a homomorphism from H into Aut(G).

The benefits to the students are manifold. Students have the opportunity to study a topic that interests them. They develop independent research skills or the ability to work in a group, depending upon the mentor's focus. By explaining their findings, students improve their ability to communicate mathematical ideas. At the completion of the undertaking, students have a sense of accomplishment. Even the weaker students feel that they have done something and take pride in their work. There are many opportunities to share the results of the student projects, either in class or outside of class.

Let us now consider examples of student accomplishments that are the result of readings, writings, and projects from two different upper division courses.

Putting it all Together: An Example from a Number Theory Course

An elementary course on Number Theory typically covers standard topics such as the number of divisors of an integer, the sum of the divisors of an integer, Euler's Φ function, congruences, ... However, specialized topics that are not covered in the lectures can become a source of reading material and projects for students.

As with all textbooks, texts in the elementary theory of numbers generally contain suggested readings. The list is compiled before the book is produced, so the references are often dated. Look for current articles for students to read in the journals listed above. A sample collection of readings for an elementary course (freshman or sophomore level) in Number Theory is provided below. For an example of a readily accessible reading, consider "September 11th Did Not Change Cryptography Policy" by Whitfield Diffie and Susan Landau. (See below.) An example of a more sophisticated reading would be "The Mobius Inversion Formula" in Vanden Eynden's *Elementary Number Theory* [5].

E. Berkovich, "A Diophantine Equation," *Pi Mu Epsilon Journal* 10 (1995) 104–10.

D. A. Cox, "Introduction to Fermat's Last Theorem," *The American Mathematical Monthly* 101 (1994) 3–14.

M. Dalezman, "From 30 to 60 is Not Twice as Hard," *Mathematics Magazine* 73 (2000) 151–53.

K. Devlin, "World's Largest Prime," *Focus* 77 (December, 1997) 1.

W. Diffie, and S. Landau, "September 11th Did Not Change Cryptography Policy," *Notices of the American Mathematical Society* 49 (2002) 448–463.

C. W. Dodge, "What is a Proof?" *Pi Mu Epsilon Journal* 10 (1998) 725–727.

J. S. Gallian, "The Mathematics of Identification Numbers," *The College Mathematics Journal* 22 (1991) 194–202.

J. Goldman, *The Queen of Mathematics: A Historically Motivated Guide to Number Theory,* A. K. Peters, Wellesley, Mass, 1998.

R. K. Guy, "Nothing's New in Number Theory?" *The American Mathematical Monthly* 105 (1998) 951–954.

B. Hayes, "The Magic Words are Squeamish Ossifrage," *American Scientist*, 82 (1994) 312–316.

L. Lamport, "How to Write a Proof," *The American Mathematical Monthly* 102 (1995) 600–608.

P. Plummer, "Divisibility tests for primes greater than 5," *Pi Mu Epsilon Journal* 10 (1995) 96–98.

E. Snapper, "What is Mathematics?" *American Mathematical Monthly*, 86 (1979) 551–557.

H. Waldman, "Tom Lehrer, Mathematician and Musician," *Math Horizons* 5 (1997) 13–15.

Topics in Number Theory are readily available for students to research. Examples include continued fractions, abstract multiplicative functions, the Möbius Inversion Formula, Pell's Equation, or applications such as check digits and cryptography. Each of these topics is suitable for a group project. Here are two examples of students' interests from the author's experience.

1. One student's fascination with magic squares led to the study of problems in nonlinear Diophantine equations.

2. Another project, entitled "Generalizations of the Jailer Problem" was the product of two students' efforts working together. Here is their description of the project:

We solved two generalizations of a famous problem. The original problem is this. All the cells in a long cell-block are locked. A first jailer turns every lock, so that they are now all unlocked. A second jailer then turns locks 2, 4, 6,.... Then a third jailer turns locks 3, 6, 9,..., a fourth turns locks, 4, 8, 12,..., etc. In the end, which cells are locked and which unlocked? In the two generalizations we solved, the nth jailer turned every $2n$th lock, and every $(2n + 1)$st lock, respectively.

Presentations of results by these students were made at Mathematics Awareness Day, held annually on the campus of Saint Joseph's University.

Finally, not all projects need to be one semester in duration. A reasonably small project would involve asking students to create their own word problems, particularly ones that are solvable in integers using Diophantine equations or the Chinese remainder theorem. Two examples from Number Theory are given below and [2] contains some additional ideas.

1. On the second of April, a zookeeper fed a llama at 6:00 AM, a dolphin at 8:00 AM and a platypus at noon. The zookeeper feeds the llama every seven hours. He feeds the dolphin every 11 hours. Lastly, he feeds the platypus every 13 hours. When the feeding times coincide, he needs assistance feeding the animals. How often does he have to find two additional zookeepers to help him feed the animals since he can't be in three places at once? When is the first day after April 2 that this happens? [Submitted by Brian Klint.]

2. Your little brother acts like a werewolf on the first day of every full moon. He has been doing this ever since he saw the movie "American Werewolf in Paris." The next full moon is tomorrow and they happen every 29 days. Your great aunt enjoys celestial events, celebrating every solstice and every equinox with a huge party in the backyard. One or the other occurs every 93 days, the next one being 18 days from now. Your mother is very devoted to her deceased parents and visits the cemetery every

Saturday. The next Saturday is four days from now. You are leaving for college exactly three years from today, but you believe you will not be able to make it if all three events happen on the same day. Do you have anything to worry about? [Adapted from an example submitted by Paul Grow.]

Puting it all Together in Probability and Statistics

A probability and statistics course is another natural place to introduce reading, writing, and projects. Data abounds in news articles, in other textbooks, in sports, and in almost every other corner of life. Reading and analyzing the numbers presented sometimes shows that the conclusions go beyond the scope of the data.

When students take a serious look at popular statistics, they begin to get a sense of the necessity for accuracy in reporting. Written explanations of their analyses are not only thought provoking but also reinforce the concepts the instructor is developing in the course. These include selecting proper sample sizes, using random sampling techniques, and the like. Below is a brief reading list for Probability and Statistics.

M. Bonsall, "Prediction in insurance," *Pi Mu Epsilon Journal* 10 (1995) 191–193.

V. Bronstein, and A. S. Fraenkel. "On a Curious Property of Counting Sequences," *The American Mathematical Monthly* 101 (1994) 560–562.

Chance (A quarterly journal), Springer-Verlag, Inc., New York.

D. Fallis, "Mathematical Proof and the Reliability of DNA Evidence," *The American Mathematical Monthly* 103 (1996) 491–497.

D. Fowler, "The Binomial Coefficient Function," *The American Mathematical Monthly* 103 (1996) 1–17.

B. Hayes, "Randomness as a resource," *American Scientist* 89 (2001) 300–305.

——, "Statistics of deadly quarrels," *American Scientist* 90 (2002) 10–15.

H. B. Hopfenberg, "Why Wars Are Lost" *American Scientist* 84 (1996) 102–104.

M. Kline, *Mathematics in Western Culture,* Oxford University Press, London, 1990.

D. P. Minassian, "The Current State of Actuarial Science," *The American Mathematical Monthly* 103 (1996) 552–561.

M. Schilling, "Things Aren't Always What They Seem," *Math Horizons* 9 (2001) 23–25.

G. Slade, "Random Walks," *American Scientist* 84 (1996) 146–153.

Stats (A biannual journal), The American Statistical Association, Washington, DC.

D. Wheeler, "The Statistics of Shape," *Math Horizons* 4 (1996) 26–28.

In a year-long course in probability and statistics, there is sufficient time for students to engage in projects at different levels. Early in the course, they can work on projects involving descriptive statistics. For instance, a team of students can compare the nutritional levels of a particular food product provided by a number of companies. A simple descriptive study can describe the fat, cholesterol, salt, protein, calories, and vitamins in a typical junk food product such as potato chips. While gaining more depth in the subject, students can think about major projects that require a much deeper analysis.

Begin preparing for major projects by discussing with each student her/his interests, career plans, hobbies, etc. Based upon this interview, projects can be found that will be challenging and interesting to the students. The process of finding suitable projects is time-consuming but well worth the effort. The instructor may be able to capitalize on the data that students collect as a hobby. This is why it is important to try to discover student interests before suggesting projects. For example, one of the author's students worked part-time with a paramedic unit. He was interested in the effectiveness of emergency paramedic care in the city of Philadelphia. The student collected the dispatch data and performed an excellent analysis. Another student had been collecting data on the local weather for a number of years and used it to analyze seasonal variations and to look for a global warming trend. Some students were interested in the relationship between a college's religious affiliation and students' views on abortion, capital punishment, and euthanasia. Finally, comparing on-line prices with those from stores and catalogs provides another topic of interest to students.

Scientists are interested in the specifications of their equipment and in the accuracy of measuring devices, such as balances and scales. Students can ascertain if the variation in measurements taken at different temperatures and different humidity levels are within the range given in the manufacturer's specifications for the instrument. Faculty members from other departments often have projects involving data collection and analysis. Mathematics students can provide a service to these professors and to their departments. For a fuller discussion of topics in probability and statistics, see [1].

1.2.4 Showcasing Student Accomplishments

A day set aside in class for students to present the results of their research is very worthwhile. If the instructor does not want to sacrifice a large block of time, smaller blocks can be used each week. Beyond the class expositions, however, there are other opportunities that allow the students to demonstrate their independent work to a broader audience. The events below describe settings that feature student presentations at all levels.

If the student project culminates in a major result, the student may be invited to be the guest speaker at the departmental honors society induction ceremony. Similarly, the mathematics department may have an honors program through which advanced students engage in independent research and expound on their results in a departmental colloquium. Most colleges and universities have an honors program in which the results of independent student research are presented to the honors faculty. Finally, your outstanding student may become a speaker at the **Pi Mu Epsilon** annual meeting held in conjunction with the summer MathFest meeting of the Mathematical Association of America. Students also may present posters at the joint annual AMS/MAA meeting in January; note that this venue has a limited group of participants and a limited audience.

All of these options apply to the brightest students. Other venues are needed so that all the students have an opportunity to demonstrate their work. For example, each year, the Association of Women in Mathematics, in conjunction with the National Security Agency, sponsors a Sonia Kovalesky Mathematics Day for high school girls. Several colleges and universities around the United States participate by holding an event on campus. Many institutions also have open houses for prospective students. Consider having the students present their projects at these events for high school students. Seeing the accomplishments of young adults is an inspiration to high school students and also a rewarding experience for the mathematics majors. At Saint Joseph's University, one activity on Sonia Kovalesky Day is a problem-solving contest. Monitored by a faculty member, undergraduate mathematics majors create the problems, moderate the contest, grade the results, and present awards to the winners. The problems are varied and interesting and usually involve only elementary mathematics. However, deductive reasoning and the ability to analyze the problem are crucial to their solutions.

Each April is Mathematics Awareness Month, sponsored by the Joint Policy Board for Mathematics [3]. The National Council of Teachers of Mathematics also sponsors a Mathematics Awareness Week. One opportunity to participate is to schedule a **Mathematics Awareness Day** in the department. As an example, consider the day's program at Saint Joseph's University. At this institution, many instructors strongly encourage their students to present a poster session for the event. Students from sophomore, junior and senior years usually make the presentations. The Pi Mu Epsilon advisor schedules talks throughout the day and coordinates the event. Students may invite their friends and the Department invites interested faculty from other departments and outside guests to attend. High school students who have been accepted as mathematics majors also may be invited to attend the activities. Although presenting a poster or project is not mandatory, student involvement is high. Some of the projects are completed jointly, while others have a single author. Presentations generally involve a poster and a verbal exposition.

Scheduled when most students are not in class, Mathematics Awareness Day begins in late morning on a Friday. Student talks are scheduled to last between 15 and 20 minutes, depending on the time needed by the presenters. In the late afternoon, a guest speaker addresses the students and faculty, with activities usually ending by 6:00 PM. Examples of some student topics for the past few years can be found on the web page

```
http://www.macs.sju.edu
```

under the heading of Students. Following the presentations, the posters are displayed in the Department area for a few days. Publicizing the event and noting that food will be served encourages students to attend, even if they are not speakers. (Serving food is always a good idea when trying to attract students.)

Student presentations can also be made in cooperation with larger organizations that include other departments. For example, the scientific research association Sigma Xi encourages research activities in its chapters. If there is an active Sigma Xi group on campus, the chapter may sponsor lectures and an annual research symposium or poster session. While generally thought of as a way for science students to demonstrate their research, mathematics majors also can display posters and explain their undergraduate research in this forum. Each spring, a regional student research symposium is held for the Philadelphia metropolitan area at Saint Joseph's University. Students from many colleges in the Philadelphia area display posters at the symposium. In 2002 there were 120 posters registered from more than twenty-five colleges and universities from five states. Areas of research included psychology, language development, mathematics, computer science, biology, physics, astronomy, engineering, chemistry, and environmental science. The abstract booklet for this Sigma Xi event is available at

```
http://www.sju.edu/honor-society/sigma-xi/book.pdf.
```

Similarly, the College of Arts and Sciences sponsors a **Scholarship Day** to exhibit the undergraduate research of all students in the College of Arts and Sciences. Once again, the best projects completed by mathematics students are presented here.

It is important to reward the students for their efforts by allowing them to show other mathematics students and faculty members their accomplishments. Students take great pride in their work. Although the depth and sophistication of the projects will vary with the talent of the students, all can participate at some level. Professors will find that once students have participated, they will want to continue to do so in future years and will have a greatly increased interest in mathematics. It is gratifying to see the students mature as their expositions become more polished with experience. The smiles you see here are genuine—students are happy displaying their results. When asked if they will participate next year, the response is an enthusiastic "Definitely!"

1.2.5 Other Creative Endeavors

The examples listed above are intended to whet the appetite of the reader. There are many other topics for mathematics faculty members to consider. We should always be inventive; that is, we should think creatively and be open to student ideas. There are many ideas for reading, writing, and creative projects which can be completed by students. Be imaginative and capitalize on the students' creativity to find projects for upper division courses. Try a small project, or simply a reading list, to get started. Look for local problems. Usually one will be able to find a few campus problems (for example, problems involving queuing theory) to analyze. And interesting problems abound. We shouldn't be afraid to learn with our students. Faculty members who accept the challenge of involving undergraduate students in reading, writing, and projects will be rewarded tenfold for their efforts.

Some of the excellent suggestions for engaging students in projects and problems arise from the efforts of faculty members from diverse campuses. Many ideas were presented during a session at the joint MAA/AMS meeting in January, 2003.[4] For example: David Knellinger (United States Military Academy) discussed "How to Solve the 'Little Things' in Life"; Murray Siegal (Sam Houston State University) explained "Using a Simulation to Model a Queuing Problem"; Bruce Pollack-Johnson (Villanova University), author of a text on the subject, explained "Teaching Modeling with Semester-Long Student-Generated Projects"; Therese Shelton (Southwestern University) introduced several modeling applications, including repeated drug doses and a room full of ping pong balls for elementary courses; and Steven Hetzler (Salisbury University) discussed "Using the Writing Process to Help teach the Mathematical Modeling Process."

For sample reading lists for a variety of courses, see the web page

$$\text{www.sju.edu/}\sim\text{arash}.$$

For suggestions for student activities and presentations, see the web site of the Mathematics and Computer Science Department at Saint Joseph's University: www.macs.sju.edu cited above. Many activities can be found under "Students." These include Mathematics Awareness Day, Sigma Xi poster sessions, and the Sonia Kovalesky Mathematics Day.

1.2.6 Conclusion

There are many ways to increase student involvement. Some of them are restricted to a particular course while others involve reaching out to the broader mathematical community. In either case, students broaden their horizons and expand their knowledge. Active learning is the key to success for any of these suggestions or programs. Whether presenting their findings in a written paper, oral presentation, or poster presentation, students enjoy the experience and take pride in their work. As mentioned in the introduction, many projects discussed lend themselves to capstone experiences. The entire department, both faculty and students, benefits from student presentations if they are displayed for all to see. There are numerous faculty members at various institutions working with students to enhance learning through readings, projects and other creative endeavors. Consult the MAA index of conferences or the MAA web site to find out more about the fascinating extensions of a variety of courses.

REFERENCES

1. A.M. Rash, "Student Projects Make Statistics Come Alive for Mathematics Majors," *Primus* VIII (1998) 193–202.
2. ———, "An Alternative Method of Assessment: Using Student Created Problems," *Primus* VII (1997) 89–95.
3. Joint Policy Board for Mathematics; consists of the American Mathematical Society, the Mathematical Association of America, and the Society for Industrial and Applied Mathematics. More information can be obtained at www.mathforum.org/mam/02.
4. *Mathematical Modeling In and Out of the Classroom,* Abstracts of Papers Presented to the American Mathematical Society, 24(2003) 238–241.
5. C. Vanden Eynden, *Elementary Number Theory*, McGraw Hill, New York, 2001.

Brief Biographical Sketch

Agnes Rash is Professor and Chairperson of Mathematics and Computer Science at Saint Joseph's University in Philadelphia, PA. She received her PhD from the University of Pennsylvania. Her mathematical interests include mathematical modeling, calculus reform, and interdisciplinary courses.

1.3

How to Develop an ILAP

Michael Huber and Joseph Myers
United States Military Academy

1.3.1 Introduction

In this guide we briefly explain what an Interdisciplinary Lively Application Project (referred to hereafter as an ILAP) is, how ILAPs are developed and executed, and what considerations and strategies arise when developing and using ILAPs. While there are many perspectives and elements to consider, we include only the essentials here and leave the rest of the material for future articles.

An ILAP is a process that generates a product that drives a student learning experience. ILAPs are student group projects that are jointly authored by a faculty member from the Mathematical Sciences Department and a faculty member from a partner department. ILAPs can be used in the mathematics classroom, in the partner classroom, or in both to let students work on mathematical concepts within the context of another discipline. ILAPs help connect the curricula by taking applications and current methods from a using department and connecting them with the concepts and techniques in the mathematics curriculum. They also can be used to reach forward to preview ideas from applications that wait downstream or backward to connect current mathematical topics with ideas from applications that already have been studied.

ILAPs provide students with practice in the interdisciplinary threads of modeling in scenarios more realistic than those usually presented within the mathematics curriculum. Students engage in reasoning (within an applied context) and problem solving, use technology as a tool to enable analysis of complex situations, connect and integrate ideas from different curricula, engage in teamwork in problem solving, and learn how to communicate methods, conclusions, and recommendations. All of this is done either in written technical reports or in a technical briefing given by the group.

The product can be formatted as the instructor desires, but experience shows that the following elements are useful:

- Problem statement.

- Background material. This is often as important for mathematics instructors as it is for students, since the application discipline is often outside their expertise.

- Sample solution. Not intended to be used as a grading guide, but rather to boost the instructors' comfort level and understanding of the problem.

- Report/briefing guidance for student authors/presenters.

There are several benefits gained by using ILAPs. These projects motivate student efforts through relevance. By showing students how current mathematical ideas are used in other disciplines, we demonstrate the areas of learning that become accessible to them as they learn and master mathematics. ILAPs seek to support student growth in the interdisciplinary threads of modeling, reasoning and problem solving, technology, and communication. ILAPs give students early experience in solving problems as part of a team. Finally, ILAPs develop partnerships among faculty that lead to discussing and developing curricula.

After you have used ILAPs for a while, you realize that *the most valuable part of ILAPs is the process of teaming and working with a partner department rather than the product that is created at the end*. We have authored many ILAPs at USMA and initially thought that we only needed to develop a few that we could then refine and recycle over a several-year timeframe . However, we have found that most of the value lies in the process rather than the product. We typically use each ILAP only once or twice, spending our time partnering and developing new ILAPs rather than refining, reusing, and publishing old ones.

Developing and Executing ILAPs

A specific Physics ILAP involved modeling the compound bow, a relatively recent improvement of the old longbow. Located at opposite ends of the bow arms are cams which allow a decrease in the amount of force required to hold the drawstring at a full draw. Students estimated an arrow's initial velocity theoretically (using a Java applet on a course webpage) and then experimentally from given data. They then calculated and discussed the uncertainty of their solutions. Maximum effective ranges were determined, which took into account certain drag forces. These requirements motivated the students to discuss all aspects of the problem using physics and mathematics. They asked questions like: Suppose the draw length or sighting angle is changed? Are there any aerodynamic forces which can be ignored? What assumptions need to be made to apply physics principles in the analysis? The scope of this example is typical of most ILAPs.

The usual chronology for developing an ILAP begins by deciding on the topic and on the skills to be used. These usually are current topics and skills from the mathematics curriculum. Then proceed as follows:

- Approach a faculty member in another department with what you want to do and ask for his or her ideas on an application that uses these skills. This involves teamwork, communication, and creativity on the part of the faculty, the same habits we seek to develop in our students. One of the important goals of the ILAP process is to gradually develop an interdisciplinary culture where partner faculty take the initiative in seeking out mathematics faculty to develop ILAPs covering concepts and skills needed in the partner discipline's courses. It is unlikely that this will happen immediately.

- Jointly write the ILAP with the partner department. Incorporate the mathematical topics and skills that you want exercised and include the partner scenarios, ideas, and connections that will be seen again by students in the partner discipline. This can be either a true joint process where both authors sit down and work together or an iterative process where one author (from either discipline) begins sketching the product, which is then refined through a series of iterations/input/revisions between the authors.

- Distribute the student handout and make the initial presentation to students. This usually involves prompting the students for what they already know about the scenario, helping them identify assumptions, seeing if they have some initial guesses about what a reasonable resolution would look like, etc. This initial presentation is often done by the mathematics faculty but it can be done with great effect by the partner faculty. It is a powerful message to students when the partner faculty visits the mathematics classroom, explains the scenario from their discipline, and challenges the students to learn the mathematics that will enable them to begin to succeed in the partner discipline.

1.3 How to Develop an ILAP

- Work begins on the ILAP. Students typically have a week or two between the time the ILAP is assigned and when it is due. We find that some degree of student-faculty interaction on the ILAP is beneficial during this period. An effective design technique is to time the ILAP so that students are learning the skills in class at the same time that they need to use them on the ILAP. Instructors will typically avoid working on portions of the ILAP in the intervening classes. They instead will work on other problems that use these concepts and will explicitly ask students how they think the material they are currently covering could be used in the ongoing ILAP. However, be careful not to let the students divide-the-work and submit or wait until the last minute to begin working on the project itself.

- Student groups make written and/or oral presentations of their analyses and solutions. Written presentations provide an excellent opportunity to develop skills in writing technical reports, with executive summaries, assumptions, analysis, conclusions, and supporting technical appendices. Students appreciate regular oral presentations as well. These presentations give them the opportunity to express themselves in person, allow instructors to ask clarifying questions on the spot rather than being forced to rule on written vagaries, and permit the team members to work together in a more integrated and balanced fashion.

After student submissions or presentations, the partner faculty can give an expert critique and extension. This helps bring closure to the project and shows how the basic ideas and skills just used are amplified and made more sophisticated in the downstream discipline. It also reinforces the work the students have done in preparing themselves to study and understand another discipline.

1.3.2 Strategy for Using ILAPs

Experience has shown us that ILAPs can be made more useful and successful by consciously using several strategies.

- Keep the application sufficiently understandable and comfortable for faculty. But it is acceptable and even healthy to get a bit beyond faculty expertise. (This reminds the faculty of what it is like to be a student.)

- Involve the students. Keep the application and the concepts involved at their level so they retain responsibility. Otherwise, students will rely heavily on assistance from instructors, peers majoring in the area of application, or other sources rather than struggling within their group.

- Capitalize on students' intuition and increase their ability to verify it. Do this by choosing scenarios involving motion, dollars, volumes, or other physical quantities so that students can apply a reality check on their own. This is much better than using quantities for which students have little feeling.

- Incorporate technology. There are a lot of complex problems out there that will be closer to what practitioners deal with. Seek to leave some of the complexity in the scenario and let students use technology to deal with it.

- Increase and intensify faculty cooperation. This is one of the main reasons for doing ILAPs, so don't try to cut it out of the process by authoring most or all of it within one department.

- Create scenarios that are flexible, multifaceted, and open-ended. The best projects come from scenarios that can be analyzed in a variety of different ways and which involve several very different constraints or considerations. Encourage multiple and, where possible, ingenious approaches to the ILAP through graphical, numerical and analytical techniques. Force students to reach beyond the algorithms and to decide how to appropriately analyze the problem they are facing.

- Be sure to include recommended guidance for students who need it. It is easy to be ambitious with ILAPs. Think about what your goals are and what you realistically expect your students to do. If you find your expectations are too ambitious, including some background in the application and provide guidance or hints on what you expect in order to make the ILAP more realistic.

Producing ILAPs which admit the need for, or actually require, student research, student discovery, and student teamwork promotes student development in these areas in ways that just are not possible with traditional activities. Since individual responsibility is sometimes neglected in group projects, consider reinforcing it with a short ILAP quiz or a question on a regular test. This will reward those who understand the key concepts and conclusions from the ILAP. In addition, encourage discussions and properly cited collaborative work between groups and others. This process introduces our students to the standards and habits of scholarship.

Focusing on one or at most a few basic mathematical concepts within the ILAP helps students understand the connection between that concept and how it arises in an application. Introducing too many concepts can prevent students from seeing this connection. At the same time, it is important to identify connections among the applications and disciplines. One of the advantages of a mathematics curriculum is that it distills ideas from several very different disciplines into a single concept. Reverse the process for students by describing how a concept that is realized in a certain way in one discipline can have different realizations in other disciplines.

The ILAPs themselves should be self-contained but at the same time be open to use in the partner discipline. ILAPs are not intended to be a feature only of the mathematics curriculum. Make them more generally useful and they will work double-duty in another department either as a starting point for discussing the application or as a way of refreshing students' memories of the mathematical concepts involved. This helps cultivate student learning from another perspective.

1.3.3 Considerations when Developing an ILAP

After developing many ILAPs for different levels of students in a variety of courses and with many different partner departments, we have found the following checklist of considerations to be useful during development:

- What mathematical topic and skills should be included? Is it a mathematical topic that needs some real life application? Or should it involve mathematical skills used in another discipline that students find hard to understand?

- What discipline or application should be our main focus? Should we play to student interest or intuition? Should we cater to mathematics faculty comfort or interests? Or should we focus on partner faculty interest and the potential for revisiting the application later in the partner department's curriculum?

- What scenario should we choose to develop? Again, do we play to student interest or to their intuition?

- How sophisticated should we make the scenario? How much information about the model should the instructor provide to the student and how much should the student discover on her/his own? Should the requirements be more prescriptive or more open-ended. Or should they be some sort of combination of both?

- What should the range of difficulty be? If easier, then maybe we should design something that can be done mostly analytically and by hand. If harder, then maybe we should make it clear that the most appropriate information we need to make a decision may be numerical, visual, and/or analytic and that it will require technology to generate and analyze.

- How do we develop group responsibility? Do we make requirements unique among the different student groups? It is valuable to have all groups working on the same scenario and requirements, especially since students report learning a lot from others when they all have a common task. But to encourage group accountability and ownership, it helps to give each group a unique set of parameter values. We have done this for ILAPs in courses of up to 1000 students, and have been pleased with the balance between collaborative learning and group accountability and ownership.

- How do we develop individual accountability of team members? Even within assigned groups, you want to see a balance between collaborative effort and individual contribution to the project. One common problem that occurs within assigned groups is that one or two members do all the work while the others do little. Or you might encounter a divide-the-work and submit approach where each member does a portion in isolation and the results are stapled together and submitted with no collaboration and no member having thought about the big picture of the project as a whole. Options available to deal with this problem include giving a follow-up quiz on the project, putting a simplified project requirement on the midterm or final, or opting for oral student presentations and spreading the questions among all group members. Of course, the act of advertising to students that these things will be done is effective in encouraging individual accountability within each group.

- Do we want to develop written presentation skills via a technical report or do we want to develop oral presentation skills via a group decision brief? Do we want to give the partner department a chance to deliver an expert critique to students, along with a summary of what students should have found. Should this critique include extensions to show how this application is treated in the partner discipline?

- How do we design the student time commitments? It is very important to know about how much student time an ILAP will require. Experience and instructor solutions help here. We figure on a ratio of 1 instructor hour equating to 3 student hours. We have found about 8 or 9 hours of work for each student to be a good requirement. Do we designate this time in the syllabus? Students need quality time to do a quality job, so rather than poach on their time we build the expected amount of student time required into the syllabus. This is usually done with class drops or reduced assignments.

- How do we grade the ILAP? What does the instructor expect of the students? Know what results you expect before you assign a project. Different levels of project difficulty can be accommodated in how you grade, as long as you know what you expect of your students. Should there be some type of standardized grade sheet? Some faculty find it useful to prepare a sheet indicating what they expect from students' final products. Grading against this type of cut sheet explicitly tells students what is expected and valued and helps standardize the grading in a large course. Others prefer to grade against their mental expectations. This is certainly quicker since it is easier for instructors to adjust expectations based on what most students are doing. This approach also allows for more judgment.

1.3.4 Guidance for Students on Written Reports

A written report is a great way for students to communicate what they have done and what they have discovered from their projects. However, effective report writing is not easy and students need help in developing their skills through practice, experience, and feedback. As students progress through their programs, most of them will be required to do technical writing in other courses. To the extent that faculty can coordinate their technical writing needs and expectations for students, we can help students grow and mature as problem solvers who can effectively communicate their analyses and recommendations to others for adoption and action.

There is no best format guide, but as an example we summarize the writing guidance that we give our students. This format fits our needs and is consistent with the format used in the courses of our partners.

Executive Summary: A one or two page summary of the scenario, including what questions were addressed, how they were addressed, and what the students' conclusions and recommendations are. This summary may be formatted as a letter addressed to the client or user if that is appropriate.

Problem Statement: This is a concise summary of the issues of interest in the given scenario.

Facts Bearing on the Problem: These are facts that are known from either the problem statement or which are uncovered during the research and problem solving stages.

Assumptions: These fill the gaps between what is known and what is required to do a successful analysis. Each must be necessary and not provable from known facts. Justification should be provided for each assumption and, if at all possible, the assumptions should be checked for consistency at the end of the analysis.

Analysis: This represents the heart of the work. It may include the following sections as appropriate: Definition of Variables and Symbols, Methodology Used, Formulas Used, Calculations, Essential Graphs and Diagrams, and Discussion of Result. (Note: Many of these sections are combined into the analysis narrative rather than separated into sections.) It is critical that the analysis be presented in narrative form, with equations and graphs used to clarify the exposition. Students are not allowed to present long multi-line derivations with no explanatory text in this section. Long derivations or supporting work that disrupts the narrative are referenced here and presented in an appendix (see below).

Conclusions and Recommendations: This presentation must follow logically from the analysis narrative. No new material should be introduced here. The contents must directly address the issues of interest from the scenario.

Appendices: As stated above, long derivations or supporting work that disrupts the analysis narrative are referenced in the narrative and presented in an appendix. Students know that each appendix must be referenced somewhere in the main body.

Acknowledgments: All sources outside the group that are used in the project must be given due credit here. We expect some amount of learning to happen from other groups, but each group must give specific credit to the group, paper, or other source from which they receive assistance. They also must be specific about what assistance was received from each source. Normal and healthy assistance is encouraged and is not penalized, while excessive reliance on a single source in a particular part of the project is discouraged and may result in a lower academic grade.

1.3.5 Grading

As our instructors assign grades to reports, they use the following institutional standards for writing. Substance is the key area we grade on, but instructors make corrections from all four areas on student projects, and can and do make grade adjustments for strengths or weaknesses they note in all areas. You may want to incorporate similar standards from your institution in your grading process. Our four writing areas are:

Substance: The correctness, completeness, and persuasiveness of the exposition.

Organization: The logical flow of the report. The format guide presented above, or a suitable student variant, is usually a big help in this area.

Style: Avoid slang, undefined acronyms, undefined or inappropriate technical jargon, and excessive use of the passive voice. Style is important in technical writing and achieving the right balance between discourse and technical expression is key to communicating with the reader.

Correctness: The document must be free of spelling, grammatical, and punctuation errors. (The correctness of the mathematics and logic is considered under the area of Substance.)

We have an obligation to our students to provide them with experience working in small groups, to develop their skills in the use of technology, to develop their communications skills (reading, writing, and presenting), to cultivate their self-esteem and confidence as problem solvers, and to demonstrate ways that their learning connects to the rest of their curriculum. We have found ILAPs to be a great way to provide appropriate developmental experiences in each of these areas.

1.3.6 Student Feedback

As stated above, one of our obligations is to provide a sense of connection between mathematics and the students' curriculum. Connecting their different learning activities empowers students to become better problem solvers in the future. We asked several students for their feedback after completing ILAPs. The responses below are from students in freshman and sophomore mathematics classes (single and multi-variable calculus) who completed ILAPs in partnership with the Computer Science, Physics, and Economics Departments.

One Computer Science ILAP dealt with designing, implementing, and testing a solution to an integral calculus problem. The student is a Hollywood analyst assigned to a film where a stunt driver is to drive a motorcycle off of the top of a building, through a protective glass wall, across a street, and into a pool on top of another building. In the single-variable calculus class, students used integration skills to determine the critical speed of the motorcycle. In the Computer Science class, students designed a flow chart and algorithm to describe the procedure their program would use to solve the program.

We discussed the compound bow Physics ILAP earlier in this paper. Another Physics ILAP involved finding a model to predict the trajectory of a projectile fired at an elevation different than the target's elevation. To make the problem more realistic, students had to account for a no-fire ceiling which the target could not penetrate for air-safety reasons. This problem in kinematics connected physics and multivariable calculus.

The Economics ILAP placed the students in a systems analysis role for the procurement of a major vehicle and asked them to estimate a reasonable cost for producing a number of vehicles in the coming year. The Cobb-Douglas equation was used for production output and data was provided on past production inputs. Students developed and refined constraint and objective functions based upon the stated requirements. Inflation was introduced and then students conducted sensitivity analyses. This problem connected optimization to a relevant area of economics.

The sophomores worked through the jointly-authored (Math/Physics) kinematics problem in their mathematics class a few weeks before studying kinematics in their physics class. One student wrote, "I feel that the project has definitely helped me to better understand the kinematics block of Physics. The math department's efforts to link mathematics to other departments is worthwhile and beneficial."

Another stated, "Project 1 was a great help with the integrations of movement that we are currently covering in physics. Linking our math project with physics was very useful and has amply prepared me for the material that is now being covered in class." Here is another perspective: "In PH203 we have just started doing those types of problems and after doing the math project solving these types of problems is much easier and I understand the concepts involved much better. I think that the connection between math and other departments is a good idea. It allows a better understanding of the applications of physics and allows some variety in the math course."

Confidence-building is another instructor obligation. Students need to be comfortable when approaching and solving *real-world* problems. One student wrote, "Physics is often easier to visualize, so applying the math to a physics-type problem of projectiles made the math make more sense." Another stated, "When the material is covered in physics, we already have a background in the math, so it's less of an exercise in memorization than in problem solving, which is what it should be. I've also noticed that the math that we've taken so far has related to economics as well as physics." A second student echoed this sentiment with mathematics, physics, and economics: "This project allowed me to further strengthen my grasp on previously covered high school material. Since physics laws and kinematics are never changing, we need to be comfortable in the application of these principles. Material I have been taught and am learning is greatly reinforced when we do assignments such as these. This second economics based math project is great because I can see all the rules and theory I have learned in my economics class really take shape."

A student who worked through a mathematics/computer science ILAP wrote, "It is very useful for different departments to cover the same ideas. Having two departments teach the same ideas will definitely contribute to the learning process. [At first,] I did not understand how the equation was created, but now I have a better understanding of both Math and CS. By showing the different ways of solving problems I think cadets will gain a more in-depth understanding of the many factors that go into solving any problem."

Students also feel that ILAPs can broaden their communications skills. One student wrote, "The project definitely gave me a good idea about the mathematics behind classical motion. As our physics course has recently transitioned into kinematics, I feel well prepared as a result of this project. As with all math projects, I was able to practice my writing skills in the explanation required."

One final comment made by a student: "I believe that it is essential that all departments, not just math, try to link into the other departments and I felt that this project did a very good job of linking into what we are just now starting to study in physics. I most definitely gained a better understanding of the physics and how these types of problems can be tackled with the help of the project. The projects do involve a lot of work and take a considerable amount of time, but I agree and think that they are essential to truly understanding a given topic, in this case projectile motion, and being able to apply what we have learned to real world situations."

1.3.7 Conclusions

This paper has presented a thorough and detailed outline of ILAPs. ILAPs connect the curricula of two departments by bringing together the applications and current methods of a user department with the concepts and toolkit techniques of a mathematics department. The applications provide relevance to the students and offer them a chance to reinforce their knowledge in more than one class. The interdisciplinary threads of modeling, reasoning, problem-solving and communicating, coupled with the integration of technology, connect the partner departments.

References

1. D. Arney and D. Small, Editors, *Changing Core Mathematics*, Mathematical Association of America, MAA Notes # 61, Washington, DC, 2002.
2. D. Arney, W. Fox, K. Mohrmann, J. Myers, and R. West, "Core Mathematics at the United States Military Academy: Leading into the 21st Century," *Confronting the Core Curriculum*, MAA Notes # 45, Mathematical Association of America, Washington, DC, (1998) 17–29.
3. D. Arney, Editor, *Military Mathematical Modeling*, West Point, NY, 1998.
4. D. Arney, Editor, *Interdisciplinary Lively Applications Projects*, Mathematical Association of America, Washington, DC, 1997.
5. D. Campbell and J. Shupenus, "Bow and Arrow Analysis ILAP," US Military Academy, 1999.
6. Department of Mathematical Sciences, *Core Mathematics at USMA*, US Military Academy, West Point, NY, 2002.

7. D. Dudley, T. Maier, and D. Olwell, "The Cobb-Douglas Problem ILAP," US Military Academy, 1996.
8. M. Gellert and T. Williams, "The Artillery Problem ILAP," US Military Academy, 1997.
9. J. Grubbs, M. Kelley, and J. Myers, "Engineering Babies in Mathematical Bathwater: the Hydrogeology ILAP," *Proceedings of the American Society of Engineering Education*, ASEE Session 3251, June 1997.
10. M. Huber and J. Myers, "Lying Truths ILAP," US Military Academy, 2002.
11. M. Kelley and J. Myers, "Diffusing Math and Environmental Engineering Through Their Common Boundary with an Interdisciplinary Lively Applications Project," *Proceedings of the Tenth Annual International Conference on Technology in Collegiate Mathematics*, G. Goodell, Editor, Addison-Wesley, (1997) 304–308.
12. J. Myers, "How to Develop an ILAP," in *UMAP/ILAP Modules 2000–01; Tools for Teaching*, Consortium for Mathematics and its Applications, Lexington, MA (2001) 21–29.
13. J. Myers, "Pedagogically Effective Technology," in *Proceedings of the Interdisciplinary Workshop on Core Mathematics*, D. Arney and D. Small, Editors, West Point, NY, (1999) 107–111.

Brief Biographical Sketches

Mike Huber received his PhD from the Naval Postgraduate School and is currently an Associate Professor at the US Military Academy at West Point. He enjoys teaching freshmen students and incorporating modeling, inquiry, and technology into solving relevant, applied problems.

Joe Myers received his PhD from Harvard University and is currently an Academy Professor at the US Military Academy at West Point. He enjoys applying mathematics to problems in the physical sciences and showing students how understanding mathematics helps them connect their learning in different areas.

1.4
The Role of the History of Mathematics in Courses Beyond Calculus

Herbert E. Kasube
Bradley University

> I am sure that no subject loses more than mathematics by any attempt to disassociate it from its history.
> — James Glaishier (1848–1928)

1.4.1 Introduction

Glaishier's comment is especially true in mathematics courses beyond calculus. A mathematics course that fails to integrate the history of the subject matter presents the mathematics in skeletal form. While the skeletal structure might be instructive, the real meat and heart of the subject are missing. This paper discusses the role that the history of mathematics can play in courses beyond calculus. It will not discuss a history of mathematics course but rather the integration of historical topics into other mathematics courses.

The first question that one asks is "Why include the history?" We could answer (sarcastically) "Why not?" but that is not an adequate answer. As mentioned above, by seeing the mathematical content without its history students see only a skeletal foundation for the subject. They fail to see some of the motivation for the subject and fail to appreciate the human effort involved. They also do not gain a proper appreciation for the process of mathematical development that has occurred. We all know that most mathematics that we study today went through a metamorphosis over the years. A student seeing this development will gain a greater appreciation for how far the subject has come since its beginnings.

An entire volume could be written about the uses of history in specific mathematics courses. The purpose of this article is more limited. Hopefully it will stimulate reader interest in the history of mathematics and in perhaps including more of it in her/his teaching. It discusses several courses in which the author has used history with good results and mentions some possible sources for other courses. Finally, it provides a fairly extensive set of references that can be a good source for history material that can contribute to the overall learning of mathematics in a variety of courses.

1.4.2 Graph Theory

I have found that an upper level course in graph theory provides an excellent opportunity to introduce a good deal about the history of the subject. The principal text for the course is by Aldous and Wilson [1], and the historical text by Biggs, Lloyd, and Wilson [3] is used as a supplement. The latter selection offers students greater historical insight into the development of graph theory, since it includes numerous original papers on the subject. The students read the selections and then discuss them in class.

The first paper to read is Leonard Euler's 1736 solution to the Konigsberg Bridge Problem, entitled *Solutio problematis geometriam situs pertinentis*, which means *The solution of a problem relating to the geometry of position*. We read and discuss Euler's entire paper, including his generalization. Some source books such as that by Calinger [9] leave out the generalization from their version of this paper. It is entirely appropriate that Euler's paper is the first one encountered by the students since it is considered the first paper ever written on graph theory.

Euler's arguments in this paper are combinatorial in nature and no modern graph theoretic terminology is used. Therefore, students can read and appreciate it before they know anything about graph theory. They quickly see that mathematical ideas are not always born in their finished form. The concept of a graph evolved through the year. Part of the beauty of Euler's work is that the exposition is clear and the steps are well motivated. Reading this paper gives the students insight into Euler's problem solving skill.

Students in graph theory also read Hierholzer's 1873 paper entitled *Uber die moglichkeit, einen linienzug ohne wiederholung and ohne unterberechnung zu umfahren* or *On the possibility of traversing a line-system without repetition or discontinuity*. This paper presents the converse of Euler's result. Students see that while Euler gave a necessary condition for a solution, he did not provide sufficiency. This allows the instructor to point out the important distinction between a necessary and a sufficient condition. Additional readings include Euler's letter to Goldbach, in which we see Euler's formula for polyhedra. This offers an opportunity to illustrate the intersection between mathematical disciplines. Euler's formula is then extended to planar graphs. Kuratowski's 1930 paper on planar graphs allows students to see some relatively recent history of mathematics.

Another topic in the course that offers a wealth of interesting history is the Four Color Problem. Many interesting individuals and events appear during this period, which began with a letter from Augustus de Morgan to William Rowen Hamilton in 1852 and ended with the final resolution of the problem by Kenneth Appel and Wolfgang Haken in 1976. Students read of Kempe's unsuccessful attempt at a proof as well as some of the controversy surrounding the use of a computer by Appel and Haken. An excellent resource for the history of this problem for both faculty and students is Robin Wilson's book [34], which makes an excellent addition to the course and serves well as a foundation for further study by the students.

1.4.3 Combinatorics

Combinatorics is another course where historical topics can be integrated successfully with course content. Here the students read the first part of Pascal's posthumously published paper *Traite du triangle arithmetique*, *Treatise on the arithmetic triangle*, that appeared in 1665. One of the interesting aspects of this section of Pascal's work is that it contains little, if any, motivation for the construction of the triangle. In an internalist's view of the history of mathematics, this is mathematics for mathematics' sake. This is in contrast with Euler's paper discussed above, which came about from an outside problem. Pascal's paper also introduces some awkward notation that can be quite confusing. This shows students that being careful with notation can be very important in helping others understand the solution to a problem. Anyone interested in learning more about the full history of Pascal's arithmetical triangle is referred to the text by Edwards [14]. In it we see that Pascal's entire paper did in fact contain applications of the triangle to both combinatorial problems and binomial expansions.

The combinatorics course also offers the opportunity to present Isaac Newton's contribution to the Binomial Theorem. In his early work with calculus, Newton looked for expansions of binomial powers

$(x + y)^{p/q}$, where p and q are positive integers. Here students see the interface between combinatorics and the calculus of infinite series, allowing them once more to recognize the connections between diverse areas of mathematics.

1.4.4 Abstract Algebra

The study of abstract algebra is rich in history, and even gives us an opportunity to see the human side of mathematics. For example, we study the youth and tragedy of Evariste Galois(1811 -1832). The fact that Galois did so much significant mathematics and died at such a young age impresses the students, who are close to his age at death. The study of his life also shows how non-mathematical events such as the French Revolution can affect the history of mathematics. A historical novel, *The French Mathematician*, written by Tom Petsinis [22], chronicles the story of Galois in some detail.

The story of Emmy Noether (1882–1935) provides another example of how world history influences the history and development of mathematics. Upon receiving her doctorate in mathematics from Erlangen in 1907, Noether found it difficult to secure an academic position. With the support of David Hilbert, she eventually came to Gottingen. Initially forbidden to teach because of her sex, Hilbert arranged for her to teach one of his courses, paving the way for a successful transition to teaching. Forced to leave Germany in 1934, Noether immigrated to the United States, where her influence on future mathematicians at Bryn Mawr cannot be exaggerated. In particular, her influence on female mathematicians was so significant that the Association for Women in Mathematics sponsors a Noetherian lecture in her honor. Noether's story is one of struggle against irrational prejudice and discrimination. Her triumph serves as an inspiration for generations to come.

We do less reading of original sources here than in the courses discussed previously. If you wish more readings, the collection of sources by Calinger [7] contains articles by Neils Abel (1802–1865), Evariste Galois (1811–1832), and William Rowen Hamilton (1805–1865).

1.4.5 Number Theory

In order to do justice to a course in number theory, you must study its history. For example, Euclid's proof of the infinitude of primes remains one of the most elegant demonstrations in all of mathematics. Virtually every modern textbook in number theory duplicates Euclid's simplicity and beauty. In addition, the study of Pythagorean mathematics introduces *Plimpton 322* as an example of Babylonian mathematics. Since this tablet predates Pythagoras by many centuries, it generates even greater interest. Recently, Eleanor Robson has discussed the origin of *Plimpton 322* in even more depth [23][24].

Pierre de Fermat is perhaps the most significant name in the history of number theory. Students are intrigued by the fact that Fermat was not a professional mathematician but rather a jurist who did mathematics for fun. Students see that a great deal of mathematics in the 17th century was done through correspondence, not in formal classroom situations. With regard to specific topics, after introducing Fermat's Little Theorem, it proves profitable to discuss Euler's generalization. This sequence illustrates how a mathematician from one century follows up on another's work from a previous one.

Arguably the most significant mathematical development of the twentieth century was Andrew Wiles' proof in 1994 of Fermat's Last Theorem. Fermat's initial conjecture has a rich history all its own. For example, Lame's incorrect proof of this result in 1847 led to the work done in ring theory by Kummer and Dedekind. Students are fascinated how a mistake could actually lead to some important mathematical results. Without the introduction of the historical context, this connection would be lost. And the way that

Andrew Wiles devoted seven years of his life to proving Fermat's Last Theorem provides a modern day illustration of mathematical dedication.

1.4.6 Other Courses

There are other courses that I have not taught that lend themselves to the use of historical material to enliven their presentations. Some of these courses and some possible historical sources are presented in the following list:

Real Analysis: *A Radical Approach to Real Analysis* [4] by David Bressoud; *Real Analysis: A Historical Approach* [28] by Saul Stahl.

Topology: *Handbook of the History of General Topology* [2] by C.E. Aull, et.al. (Ed); *History of Topology* [16] by T.M. James.

Statistics: *Statistics on the Table: The History of Statistical Concepts and Methods* [29] or *The History of Statistics: The Measurement of Uncertainty Before 1900* [30], both by Stephen Stigler; *The Lady Tasting Tea: How Statistics Revolutionized Science in the Twentieth Century* [25] by David Salsbury.

Linear Algebra: *Resources for Teaching Linear Algebra* [10] by D. Carlson et.al. (Ed)

Complex Analysis: *Higher Calculus: A History of Real and Complex Analysis from Euler to Weierstrass* [5] by Umberto Bottazini; *Cauchy and the Creation of Complex Function Theory* [27] by Frank Smithies.

1.4.7 Assessment

Assessment of historical topics within a course can be difficult. For example, it may be possible to include one or more historically oriented questions on some or all of the examinations offered during the course. Essays or term papers are another option that can help determine student understanding of the overall role of the history of mathematics in whatever area is being discussed.

1.4.8 Where do you find this stuff?

You can start by choosing a good survey text in the history of mathematics. Some examples include Burton [6], Calinger [8], Cooke [11], Grattan-Guiness [15], and Katz [17]. As for original sources in mathematics, some suggestions include Calinger [9], Smith [26], and Struik [31]. The text by Dunham [13], while not exactly a collection of original sources, studies the work on Euler in some depth. The text by Laubenbacher and Pengelley [19] offers another look at original sources. But perhaps the best suggestion is just to READ! Read a lot of varied materials; there is a great deal out there.

You can find many books about specific mathematical topics that would be most appropriate for the course in question. Dunham's book [13] on Euler already has been mentioned. Possible references for a course in probability and statistics would include the books by David [12] and Stigler [30]. Some textbooks contain more historical material than others. For example, Tattersall's recent number theory text [33] is rich in historical content and Stahl's text in abstract algebra [28] also takes the same direction. Historical content is one thing to look for when choosing a text for a course. And there is always the internet; who knows what you might find if, for example, you were to enter "history" along with some mathematical term in your favorite search engine.

1.4.9 Conclusion

Introducing historical content into the courses beyond calculus can be a great deal of work and must be done carefully. It is important that the historical topics not appear to be thrown in just to fill time. Historical content should be used to motivate what is to come or, somewhat less often, to explain what has happened. For example, in the graph theory course described above, Euler's paper was discussed very early in the course, before the students had a clear view of exactly what a graph is. After all, that is how Euler saw it. Similarly, the discussion of Pascal's paper in combinatorics takes place at the very beginning of the material on the Binomial Theorem. Students must view a subject's history as an inherent part of its study.

Student reaction to the inclusion of historical content has been very positive. Reactions have varied from simply "That's interesting!" to "It made the mathematics more interesting." In general, students seem to find courses taught this way more interesting and less dry than one using a more traditional approach.

References

1. Joan Aldous and Robin Wilson, *Graphs and Applications: An Introductory Approach*, Springer Verlag Publishing, 2000.
2. C.E. Aull et.al. (Ed), *Handbook of the History of General Topology*, Kluwer Academic Publishers, 2002.
3. Norman Biggs, Keith Lloyd, and Robin Wilson, *Graph Theory: 1736–1936*, Oxford University Press, 1998.
4. David Bressoud, *A Radical Approach to Real Analysis*, Mathematical Association of America, 1996.
5. Umberto Bottazzini, *A History of Real and Complex Analysis from Euler to Weierstrass*, Springer-Verlag, 1986.
6. David Burton, *The History of Mathematics: An Introduction* (Fourth Edition), McGraw Hill Publishing, 1999.
7. Ronald Calinger, *Classics of Mathematics*, Prentice Hall Publishing, 1995.
8. —— (ed.), *Vita Mathematica*, MAA Notes # 40, Mathematical Association of America, 1996.
9. ——, *A Contextual History of Mathematics*, Prentice Hall Publishing, 1999.
10. D. Carlson et.al. (Ed), *Resources for Teaching Linear Algebra*, MAA Notes, Volume 42, 1997.
11. Roger Cooke, *The History of Mathematics: A Brief course*, John Wiley and Sons, 1997.
12. F.N. David, *Games, Gods, and Gambling: A History of Probability and Mathematical Ideas*, Dover Publications, 1998.
13. William Dunham, *Euler: The Master of Us All*, Dolciani Mathematical Expositions, #22, Mathematical Association of America, 1999.
14. A. W. F. Edwards, *Pascal's Arithmetical Triangle: The Story of a Mathematical Idea*, Johns Hopkins University Press, 2002.
15. Ivor Grattan-Guiness, *The Rainbow of Mathematics: A History of the Mathematical Sciences*, W.W. Norton Co., 1997.
16. T. M. James, *History of Topology*, North Holland Publishing, 1999.
17. Victor Katz, *A History of Mathematics: An Introduction* (Second Edition), Addison Wesley Publishing, 1999.
18. —— (ed.), *Using History to Teach Mathematics: An International Perspective*, MAA Notes # 51, Mathematical Association of America, 2000.
19. Reinhard Laubenbacher and David Pengelley, *Mathematical Expeditions: Chronicles by the Explorers*, Springer Verlag Publishing, 1999.
20. Karen Parshall and David Rowe, *The Emergence of the American Mathematical Research Community 1876–1900, J.J. Sylvester, Felix Klein, and E.H. Moore*, American Mathematical Society, 1994.
21. Karen Parshall, *James Joseph Sylvester: Life and Work in Letters*, Oxford University Press, 1998.
22. Tom Petsinis, *The French Mathematician*, Berkley Publishing, 1997.
23. Eleanor Robson, Neither Sherlock Holmes nor Babylon: A Reassessment of Plimpton 322, *Historia Mathematica*, 28 (2001), 167–206.
24. ——, Words and Pictures: New Light on Plimpton 322, *American Mathematical Monthly*, 109 (2002), 105–120.

25. David Salsbury, *The Lady Tasting Tea: How Statistics Revolutionized Science in the Twentieth Century*, Owl Books, 2002.
26. David Smith, *A Source Book in Mathematics*, Dover Publishers, 1959.
27. Frank Smithies, *Cauchy and the Creation of Complex Function Theory*, Cambridge University Press, 1997.
28. Saul Stahl, *Introductory Modern Algebra: A Historical Approach*, John Wiley Publishing, 1997.
29. Stephen Stigler, *Statistics on the Table: The History of Statistical Concepts and Methods*, Bellnap Press, 1990.
30. ——, *The History of Statistics: The Measurement of Uncertainty Before 1900*, Harvard University Press, 1986.
31. Dirk Struik, *A Source Book in Mathematics: 1200–1800*, Princeton University Press, 1990.
32. Frank Swetz, et.al. (ed.), *Learn from the Masters*, Mathematical Association of America, 1995.
33. James Tattersall, *Elementary Number Theory in Nine Chapters*, Cambridge University Press, 1999.
34. Robin Wilson, *Four Colors Suffice: How the Map Problem was Solved*, Princeton University Press, 2002.

Brief Biographical Sketch

Herbert E. Kasube received his PhD in mathematics from the University of Montana. He currently is an associate professor of mathematics at Bradley University in Peoria, Illinois. His mathematical interests include number theory, combinatorics, and, of course, the history of mathematics.

1.5
A Proofs Course that Addresses Student Transition to Advanced Applied Mathematics Courses

Michael A. Jones and Arup Mukherjee
Montclair State University

1.5.1 Introduction

At many institutions, the standard transition for undergraduate students from the calculus sequence to upper level courses in mathematics involves a proofs course. One of its purposes is to mature undergraduate students and change their perspective from problem solving to theorem proving. In such a course, students learn about the abstract nature of mathematics while at the same time learning how to construct basic proofs, how to read mathematics, and how to write mathematics. Of course, it is impossible to teach how to prove without proving something! Proofs courses often introduce concepts and topics from a variety of mathematical fields, thereby providing a sample of advanced pure mathematics.

A survey of some recent textbooks designed for proofs courses indicates the wide variety of topics used to introduce the concept of proof. For example, Schumacher [11], Eisenberg [4], and Fletcher and Patty [5] focus on number theory, axiomatic approaches to examining the real numbers, and the cardinality of sets. Rotman [10] offers less of a sampling of higher mathematics, but grounds the proofs in mathematics more familiar to students, including geometry, trigonometry, and properties of polynomials. Of course, the treatment is much more precise and rigorous than the students may have seen and does develop and use more advanced mathematics in these more familiar areas. D'Angelo and West [16] provide a more extensive sampling of advanced mathematics, including discrete mathematics (probability, combinatorics, graph theory, and recurrence relations) and continuous mathematics (sequences, series, continuity, differentiation, and Riemann integration). All of the aforementioned texts have chapters or appendices that introduce elementary set theory, induction, the properties of functions and relations, and equivalence relations.

Despite the notion that a proofs course contains a stable of techniques used to prove different assertions, not all of the texts include chapters on proof techniques, quantifiers, logic, *etc*. There are other texts that focus on the processes of proving and writing mathematical results. These include Solow [14] and Velleman [15]. In particular, Velleman [15] breaks the process of proving a result into smaller pieces and discusses how scratch-work evolves into the final wording of a proof.

There are a number of takes on how to prepare students for upper level mathematics. Like the aforementioned texts, our proofs course introduces students with a calculus background to standard techniques of proof through different topics in mathematics. Motivated by our attendance at an NSF workshop on Project InterMath, we decided that using the transition from discrete to continuous mathematics would provide a good setting for a proofs course at the sophomore level. The mathematical topics in the course come from difference equations, differential equations, and elementary linear algebra. However, rather than consider the course as a sampler of advanced pure mathematics courses, we view our course as a sampler of the many applications of mathematics.

For our institution, we believe that the proofs course should not only enhance the ability of our students to communicate verbally and through writing but also contain a heavy technological component and exploratory aspect. We want students to be able to make conjectures and have an experience that introduces them to the process of research. As for the mathematical content of the course, we considered what materials we wanted our students to know *before* they entered our upper level courses. In this article, we not only describe the course, but also give the background of our institution and how this course was designed to solve certain problems in our program. By design, the proofs course offers a sampling of the different majors and programs offered in our department. Also, since most of the active research faculty at Montclair work in applied mathematics, the course provides an opportunity for students to learn about the research areas of faculty and to be better prepared to pursue undergraduate research.

The course ran as an elective in academic years 2003–2004 and 2004–2005. After undergoing evaluation in spring 2004, the mathematics faculty voted to support making the course required for mathematics majors as part of a six credit increase in the total number of mathematics credits taken by our majors. Although the transitions course does not have a long history at Montclair, much of the content for the course has been used by the authors in courses in Calculus I/Discrete Dynamical Systems, Game Theory, and Linear Algebra at the U.S. Military Academy at West Point, in Game Theory, Differential Equations, and Mathematical Modeling at Montclair, and in Differential Equations at Rutgers. This article includes an outline of the course and provides examples of some of the course content. We conclude by discussing how to adapt our course to another institution, as well as ruminating on the purpose of a proofs course with suggestions on how to develop a course tailored to other constraints.

1.5.2 Fitting the Proofs Course into the Institution

Montclair State University's Department of Mathematical Sciences graduates about 30 majors in mathematics every year. Approximately 35% of our majors are transfer students from other institutions, primarily from the local community or county colleges of New Jersey. A number of our majors begin by taking remedial courses in mathematics—typically one semester of Pre-Calculus.

Table 1.5.1 gives an outline of the typical mathematics courses our majors take. We have two mathematics streams—Mathematics and Applied Mathematics—the latter being further sub-divided according to specialization into Track I (Discrete Applied Mathematics and Operations Research) and Track II (Statistics). The minimum required credits in mathematics for a typical student majoring in mathematics is 40 while a student majoring in the applied mathematics concentration takes 49; c.f. Table 1.5.1. The total credit requirements for our majors is specified in Table 1.5.2. Additional mathematics courses are chosen as *free elective* requirements by the students(see last row of Table 1.5.2).

The only courses that all majors take at the freshman/sophomore level are the Calculus sequence and Linear Algebra. Many students take the Calculus courses off sequence (beginning Calculus I in the spring of their freshman year) because they take Pre-Calculus in the fall of their freshman year. These students may take Calculus III and Linear Algebra concurrently to catch up in the spring of their sophomore years. Despite the emphasis on taking these courses by the end of the sophomore year, some students take

1.5 A Proofs Course that Addresses Student Transition to Advanced Applied Mathematics Courses

All Majors (19 credits)		
Calculus I, II, III + Linear Algebra + Probability		
Mathematics (9 credits) *Foundations of Computer Science I* *Advanced Calculus I* *Foundations of Modern Algebra*	**Applied Mathematics** (9 credits) *Foundations of Computer Science I + II* *Introduction to Mathematical Modeling*	
	Specialization	
	Track I (9 credits) *Discrete Math* *Operations Research I* *Operations Research II*	**Track II** (9 credits) *Statistical Methods* *Statistical Computing* *Mathematical Statistics*
At least 4 electives (12 credits)	At least 4 electives (12 credits)	At least 4 electives (12 credits)
Major requirements (Mathematics courses)		
$(19 + 9 + 12) = 40$ credits	$(19 + 9 + 9 + 12) = 49$ credits	

Table 1.5.1. Required and elective (mathematics) courses for Mathematics and Applied Mathematics Majors at Montclair State University

Linear Algebra in the fall of their junior year while also taking more advanced courses. Montclair State University's Calculus sequence is fairly traditional with nods to calculus reform. All Calculus courses use the Larson, Hostetler, and Edwards [8] text. The department policy is for all students to have a TI-86 calculator. Students are allowed to use their calculators in lectures, on quizzes, and for exams. Faculty are expected to incorporate the graphing calculator into their lectures. Because the department does not require group work or written assignments in the Calculus sequence, student exposure to these pedagogical devices is largely instructor dependent. It is possible for students in the higher-level courses to have had no experience with group work, exploration, or writing.

From this point onward, we use the phrase *higher level courses* to refer to any course beyond the required courses of the Calculus sequence (Calculus I–III), Linear Algebra, and Probability. Because some higher level courses are offered every term while some are offered every year, and because these courses typically do not have pre-requisites other than some of the five aforementioned courses, the background of students in the higher level courses can vary dramatically. In any given course, there may be students in the last term of their undergraduate careers as well as students who are taking their first higher level courses. The proofs course decreases the degree of variability of mathematical preparation of students in higher level courses.

Some of the key points about the knowledge, preparedness, and maturity of our students are listed below:

- Almost all our students taking higher level courses have had no exposure to proofs (except in their Linear Algebra class for which some students receive transfer credit).
- Due to the way our required courses are structured, students typically do not have a chance to get abundant practice with descriptive writing projects.
- Our students have little or no experience with computer algebra systems (for example *Maple*) or spreadsheets (for example *Excel*).
- The minimal knowledge base of our students in higher level courses is limited and varies widely.
- Our students can graduate with their only exposure to differential equations being the brief reference to separable equations in calculus.
- Although students are introduced to applications in Calculus I and II, they are not exposed to a more diverse set of applications of mathematics until they enroll in higher level courses. At this time, they

already may have made decisions about what classes to take and whether or not to add a concentration in either pure or applied mathematics.

The proofs course is designed to address most of the issues mentioned above. Before the introduction of the proofs course, any student with the prerequisites of the calculus sequence and linear algebra could take any of the higher level courses. Hence, the mathematical maturity and experience of students in these courses varied greatly. The introduction of a new sophomore level prerequisite increases the minimal experience of the students in the higher level courses. Also, since a number of our transfer students take linear algebra at other institutions, the proofs course ensures that students in the higher level courses have been introduced to proofs, applications, and exploration.

Students in the proofs course are introduced to the rudiments of writing proofs early in the term. This is achieved by introducing proofs in the context of difference equations. For example, we use induction to prove the form of a general solution to a second order linear homogeneous difference equation. During this process, the students are encouraged to explore the structure of the general solution using *Maple* or *Excel*, conjecture the form of the solution, and then prove their conjecture using induction. The close interplay between exploration, conjecture, and proof forms a structure which is reinforced throughout the course as new topics are introduced.

Since Montclair does not have a sophomore-level differential equations course, some of our graduates have only the limited exposure to differential equations that they received in the calculus sequence. The proofs course gives the students a better idea of the usefulness of differential equations and the elementary solution techniques. Our course is designed to introduce students to applications of mathematics to real world problems at an early stage by using difference and differential equations as modeling tools. This course thus serves as an advertisement for careers in mathematics and for the different courses and programs in our department.

Most of our mathematics majors intend to become high school mathematics teachers. Montclair has a strong Mathematics Education group in the department, and entrance into the teacher education program for students majoring in mathematics is competitive. Many students who decide to major in mathematics because they want to teach find that they are not interested in the teacher education program. It is our belief that some of these students decide that they want to teach because they like mathematics but are unaware of other career opportunities.

By introducing a variety of applications, the proofs course demonstrates that mathematics is a tool that can be used to model and solve problems in different fields. The proofs course also provides an opportunity to introduce students to the areas of applied mathematics in which the faculty are currently active. By selecting applications that are germane to the research of the faculty, the students get to know what the faculty are doing and learn about opportunities in undergraduate research. A handful of faculty at Montclair are involving undergraduates in their research, but the common complaint is that students become involved in the research too late in their undergraduate careers.

1.5.3 Outline, Description, and Philosophy

The prerequisite for the proofs course is two semesters of calculus. The course material emphasizes exploration, applications, technology, and proofs. It combines topics from Discrete Dynamical Systems, Linear Algebra, and Differential Equations with a strong exploratory and writing component. Students are required to work in groups and turn in projects throughout the term, culminating in a capstone project at the end of the course. Many of the applications are taken from Interdisciplinary Lively Application Projects (ILAPs) by Arney [1]. ILAPs originated at the United States Military Academy at West Point through the

1.5 A Proofs Course that Addresses Student Transition to Advanced Applied Mathematics Courses

Mathematics major	Applied Mathematics major
Major requirements (mathematics courses) from Table 1	
40 credits	49 credits
Collateral course requirements (non–mathematics courses)	
8 credits	7–9 credits
General Education requirements (non–mathematics courses)	
46 credits	46–48 credits
Free elective requirements (mathematics (or other) courses)	
26 credits	14–18 credits

Table 1.5.2. Mathematics, Collateral, General Education, and Free elective requirements for Mathematics and Applied Mathematics majors at Montclair State University

interaction of the mathematics department with the science, engineering, and social science departments. ILAPs are integrated, student-centered projects linking mathematics with partner disciplines.

In addition to the group projects, students are required to conjecture results relating to the topics being covered through exploration and to write detailed and comprehensive proofs individually. Approximately half the class time is spent on exploration while the other half is dedicated to rudimentary proof techniques. Our philosophy of using exploration as a preamble to mathematical rigor and proof has been explored and used in proofs courses before. During the 1990's, Mount Holyoke [9] developed a bridge course titled *Laboratory in Mathematical Experimentation*. Similar to our transitions course, this course is taken by all mathematics majors at the beginning of their sophomore year and the prerequisite is two semesters of calculus. The course at Mount Holyoke lets students learn a wide range of topics in mathematics through discovery and experimentation. The students, working in small groups, are encouraged to explore topics, make conjectures, and then construct arguments in support of those conjectures. The preface in [9] states that

> Students who have taken the Lab course are more likely to ask questions and look for patterns, to formulate arguments clearly, and more likely to dive in and mess around with a hard problem. Moreover, students who have taken the course do better in Real Analysis and Abstract Algebra than students who have not.

The Mount Holyoke course focuses on proofs and pure mathematics through exploration. The spirit of our course is very similar to the Mount Holyoke course but our course is centered on proof and exploration in applied mathematics.

The ILAPs form a collection of engaging problems which interest and even excite the students. Our course is designed to use the students' curiosity about these applications as a starting point to help them learn the value of exploration, to guide them into forming conjectures from their explorations, and to teach them how to use mathematical reasoning to prove their conjectures. D'Angelo and West [16] mention that the inherent difference between the focus on computation in lower level courses and on attention to careful exposition in the higher level courses presents a major challenge to many students. Various proofs and transition courses have different approaches to address this challenge. These approaches are usually based on the focus and specific needs of the institutions where the courses are taught. The lack of a required differential equations course at Montclair State University and the attempt to attract more students who will be interested in applications of mathematics led us to our take on the transition course. We also schedule faculty visits throughout the term during which a faculty member has 15 to 30 minutes to loosely explain her/his research to the class. This fosters a sense of community, lets students meet other faculty members, and helps recruit students for undergraduate research projects.

We now give an outline of the specific topics covered in the course:

Weeks 1–2: Review sequences from Calculus II. Introduce Discrete Dynamical Systems. Examine first-order, linear homogeneous difference equations. Explore numerical solutions and their long-term behavior using Excel. Introduce a detailed application through an ILAP (for example, saving to buy a car).

Weeks 3–4: Conjecture solutions and introduce induction to prove conjectures. Analytically solve difference equations. Continue to apply difference equations to model real-world behavior. Assign the first project. Faculty visit #1.

A specific example on the use of induction to prove the general form of the solution to a second-order, linear homogeneous difference equation is given in the next section. The example illustrates the *exploration leading to conjecture leading to proof* structure of our course. During Weeks 2 and 3 the students explore specific examples of difference equations. In particular, they use spreadsheets (for example Excel) to study the long term behavior of solutions and to conjecture the structure of general solutions. Finally, they prove their conjecture rigorously. The proof is done carefully in class for at least one case (for example, when the characteristic equation of a second-order linear homogeneous difference equation has distinct roots), and the other cases are assigned as homework.

Weeks 5–7: Consider systems of difference equations by introducing elementary ideas from Linear Algebra. Have students explore ideas about stability, eigenvalues, and eigenvectors using *Maple*. Develop models to highlight concepts and applications.

Weeks 8–9: Use limits as a transition from difference equations to differential equations. Examine slope fields with *Maple*. Assign the second project. Faculty visit #2.

This course uses the development of models to highlight concepts. A specific example from evolutionary game theory is given in the next section. The mating strategies used by lizards in California are modeled using discrete and continuous models. This provides an opportunity to illustrate the transition and connection between difference equations, which is the main topic in the first third of the course, and differential equations, which forms the last third.

Weeks 10–12: Consider first-order and second-order differential equations with constant coefficients. Analyze a real-world application using *Maple* (for example, population models). Assign the capstone project.

Weeks 12–14: Examine systems of linear differential equations including simple concepts of stability. Compare discrete and continuous models using technology. Develop models of real-world phenomena (for example, predator-prey models). Faculty visit #3.

Week 15: Small groups report on capstone projects.

The material on fundamental solutions of linear, homogeneous differential equations in weeks 12–14 serves as an illustration of the interplay between exploration, conjecture, and proof. The material is standard and can be found in a variety of texts on differential equations such as Boyce and DiPrima [2]. We expand on this interplay in Example 3 in the next section.

1.5.4 Some Specific Mathematical Content

Mathematical induction is often the first type of proof that is taught to students. To demonstrate the use of induction, we provide the following example. Students in the proofs course use induction to prove that solutions of a particular form are general solutions to linear, homogeneous difference equations. Notice that this does not indicate that all solutions are of this form and therefore is not a proof that the general solution to the difference equation is of a particular form.

1.5 A Proofs Course that Addresses Student Transition to Advanced Applied Mathematics Courses

Example 1 Using induction to prove the form of a general solution to a second-order, linear homogeneous difference equation:

Assume that the second-order, linear homogeneous difference equation is of the form $a(n) = sa(n-1) + ta(n-2)$ where s and t are constants and a is a function of n that satisfies the given recursive relationship. The characteristic polynomial associated with the difference equation is $x^2 = sx + t$.

Suppose that the characteristic polynomial has distinct roots r_1 and r_2. Then, the general solution to the difference equation has the form $a(n) = c_1 r_1^n + c_2 r_2^n$ where c_1 and c_2 are arbitrary constants that are determined by the initial conditions.

To begin this proof, we must first establish the base case, where $a(0) = c_1 r_1^0 + c_2 r_2^0 = c_1 + c_2$ and $a(1) = c_1 r_1^1 + c_2 r_2^1 = c_1 r_1 + c_2 r_2$. Then we obtain

$$a(2) = sa(1) + ta(0) = s(c_1 r_1 + c_2 r_2) + t(c_1 + c_2)$$
$$= c_1(sr_1 + t) + c_2(sr_2 + t) = c_1 r_1^2 + c_2 r_2^2$$

where the last equality holds because r_1 and r_2 are roots of the characteristic polynomial.

To prove the general case by induction, assume that $a(n) = c_1 r_1^n + c_2 r_2^n$ for nonnegative integers less than or equal to n. By substitution, it follows that

$$a(n+1) = sa(n) + ta(n-1)$$
$$= s\left(c_1 r_1^n + c_2 r_2^n\right) + t\left(c_1 r_1^{n-1} + c_2 r_2^{n-1}\right)$$
$$= c_1 \left(sr_1^n + tr_1^{n-1}\right) + c_2 \left(sr_2^n + tr_2^{n-1}\right)$$
$$= c_1 r_1^{n+1} + c_2 r_2^{n+1},$$

where this last equality follows by substitution since r_1 and r_2 are the roots of the characteristic polynomial. Hence, by induction, if the roots are distinct, then the general solution is $a(n) = c_1 r_1^n + c_2 r_2^n$.

Suppose that the characteristic polynomial has a repeated root, r. The general solution to the difference equation has the form $a(n) = c_1 r^n + c_2 r^n n$ where c_1 and c_2 are again arbitrary constants that are determined by the initial conditions. The base case is established in the same way as it was above. To prove the general statement by induction, assume that $a(n) = c_1 r^n + c_2 r^n n$ holds for nonnegative integers less than or equal to n. Since the root is repeated, the characteristic polynomial is $(x - r)^2 = x^2 - sx - t$. This implies that $t = -r^2$ and $s = 2r$. By multiple substitutions, it follows that

$$a(n+1) = sa(n) + ta(n-1)$$
$$= s\left(c_1 r^n + c_2 r^n n\right) + t\left[c_1 r^{n-1} + c_2 r^{n-1}(n-1)\right]$$
$$= c_1 \left(sr^n + tr^{n-1}\right) + c_2 \left[sr^n n + tr^{n-1}(n-1)\right]$$
$$= c_1 \left(2r^{n+1} - r^{n+1}\right) + c_2 \left[2r^{n+1} n - r^{n+1}(n-1)\right]$$
$$= c_1 r^{n+1} + c_2 r^{n+1}(n+1).$$

Hence, by induction, if the roots are repeated, then the general solution is $a(n) = c_1 r^n + c_2 r^n n$. The entire proof is then complete.

We consider two formulations of the same biological model of the competitive mating strategies of lizards in California. This example was motivated by an article in *The Economist* [7] that in turn reported on an article in *Nature* by Sinervo and Lively [13]. A more complete comparison of the differences between the discrete and continuous approaches appears in Weibull [17] and has been successfully used in game theory courses by Jones to focus on the subtleties of modeling dynamic behavior. This example not only demonstrates the applicability of mathematics to model reality but also introduces students to areas of research (mathematical biology and game theory) in which Montclair faculty are active. A short discussion on implementation and pedagogy follow the next example.

Example 2 An application to evolutionary game theory/biology that depends on the discrete or continuous formulation:

Rock-Paper-Scissors or Roshambo is a two-player game with no clear choice of which of the three options is best to employ. Rock crushes scissors, paper covers rock, and scissors cuts paper. The play results in a tie if both of the players select the same action, e.g., rock versus rock. If players in a population only play one of the three alternatives and are randomly matched with an opponent, then the optimal play would be to play the strategy that defeats the most frequently played action. Without such frequency information, it is not surprising that the best strategy is to randomize between rock, paper, and scissors, playing each with probability $\frac{1}{3}$. What is surprising is that Rock-Scissors-Paper models the mating strategies of lizards in California. As reported in Sinervo and Lively [13], male lizards with different colored throats use different mating strategies and pass their strategies to their similarly colored male offspring. The relationship between the strategies employed by the lizards is the same as the relationship between the strategies of playing rock, paper, or scissors in Roshambo. That is, there are three strategies and each strategy becomes more successful when another strategy becomes more frequent. For a more detailed explanation of the strategies, placing them into the context of the actual practices of the lizards, consider Sinervo and Lively [13] or the website [12].

When a mating strategy becomes more successful, it results in more offspring of the lizards that employ the strategy. The evolutionary process can be modeled both discretely and continuously. We consider the discrete process first. We revert back to Rock-Paper-Scissors to discuss the evolutionary process. Let (a, b, c) where $a + b + c = 1$ and $a, b,$ and $c \geq 0$ represent the percent of the population at time n playing the game. The generation of the population at time $n + 1$ depends on the population distribution at n. How the population evolves depends on the relationship between a, b, and c. Specifically, the rock players tie against one another, lose against paper players, and win against scissors players; this occurs with probability a, b, and c respectively. A player receives 2 points for a win, 1 point for a tie, and 0 points for a loss. Mimicking evolutionary processes, the success of players using a strategy against the population at time n determines the number of offspring who use the strategy at time $n + 1$. Rock-Paper-Scissors players receive $a + 2c$, $b + 2a$, and $2b + c$ points respectively. Since the sum of all points is $a + 2c + b + 2a + 2b + c = 3(a + b + c) = 3$, the distribution of the population at time $n + 1$ is

$$\left(\frac{a + 2c}{3}, \frac{b + 2a}{3}, \frac{2b + c}{3}\right).$$

Hence, the evolutionary process can be viewed as a Markov chain or system of difference equations where the population at time t is given by $[a(n)\ b(n)\ c(n)]^T$ and the next generation can be determined by matrix multiplication:

$$\begin{bmatrix} \frac{1}{3} & 0 & \frac{2}{3} \\ \frac{2}{3} & \frac{1}{3} & 0 \\ 0 & \frac{2}{3} & \frac{1}{3} \end{bmatrix} \begin{bmatrix} a(n) \\ b(n) \\ c(n) \end{bmatrix} = \begin{bmatrix} a(n+1) \\ b(n+1) \\ c(n+1) \end{bmatrix}.$$

As long as the initial population consists of nonzero populations, the distribution tends to $\left[\frac{1}{3}\ \frac{1}{3}\ \frac{1}{3}\right]^T$. The matrix describing the evolution of the population is doubly stochastic, because all rows and columns

1.5 A Proofs Course that Addresses Student Transition to Advanced Applied Mathematics Courses

sum to 1, see, e.g., Isaacson and Madsen [6]. All doubly stochastic matrices of dimension m have the m-dimensional vector with all entries $\frac{1}{m}$ as a fixed point or equilibrium vector. This vector is unique and attracting if the matrix is also ergodic [6]. As the matrix above is ergodic, students can arrive at the equilibrium vector by iterating the matrix. Students can plot the long-term behavior using *Excel* or *Maple* and see that the vectors converge to $\begin{bmatrix} \frac{1}{3} & \frac{1}{3} & \frac{1}{3} \end{bmatrix}^T$.

The above depiction of evolution in a society of Rock-Paper-Scissors players is by finite replicator dynamics, as described in Weibull [17]. However, the evolution can be modeled continuously. Let the current population be given by $[a \ b \ c]^T$ where a, b, and c are functions of time. For $s \in \{a, b, c\}$, the growth rate $\frac{s'}{s}$ of the portion of the population using strategy s equals the difference between the strategy's current payoff and the current average payoff in the population. The rock strategy's current payoff is given by $a + 2c$; the paper strategy's current payoff is $b + 2a$; the scissors strategy's current payoff is $2b + c$. These are calculated as before. The current population $[a \ b \ c]^T$ receives on average a payoff of

$$a(a + 2c) + b(b + 2a) + c(2b + c) = a^2 + 2ac + b^2 + 2ab + 2bc + c^2 = (a + b + c)^2 = 1.$$

A little algebraic manipulation yields $\frac{a'}{a} = a + 2c - 1$ or $a' = (a + 2c - 1)a$. Similarly, for b and c the system of differential equations becomes

$$a' = (a + 2c - 1)a$$
$$b' = (b + 2a - 1)b$$
$$c' = (c + 2b - 1)c.$$

Once again, the vector $\begin{bmatrix} \frac{1}{3} & \frac{1}{3} & \frac{1}{3} \end{bmatrix}^T$ is an equilibrium or fixed point. However, unlike the discrete version, where populations converge to the fixed point, in the continuous model, the vector is neither attracting nor repelling. Indeed, it is a center and trajectories of $[a(t) \ b(t) \ c(t)]^T$ orbit around the equilibrium. Using *Maple,* students can plot the trajectory of the orbits on the simplex

$$S = \{(a, b, c) \mid a + b + c = 1; a, b, c \geq 0\}.$$

Pedagogically, the discrete version is introduced earlier in the term (see Weeks 5–7 in the outline). The data presented in Sinervo and Lively [13] and on the website [12] shows oscillatory behavior and the population does not converge to the equilibrium vector, as predicted by the discrete time model. This provides an opportunity to discuss alternate ways to model the changes in the population. The differential equation version of the lizard mating game is considered later in the term (see Weeks 12–14 in the outline). By revisiting the evolutionary model of the lizard population, students learn that there is no one way to model reality. This is also a good time to discuss how a model should be developed and compared to reality. Similarly, the comparisons between the discrete and continuous models should focus on the increased complexity of the continuous model and how simple models are valued as long as they accurately model reality.

As an illustration of the material from the last portion of the course, we now explain how exploration leads to conjecture and proof when examining linear, homogeneous, differential equations.

Example 3 Exploration, conjecture, and proof in linear, homogeneous differential equations:

Let us consider the simple second-order homogeneous equation $y'' - y = 0$ where $y = y(t)$ is the solution. Building on their knowledge of the exponential function from Calculus, the students quickly verify that $y_1(t) = e^t$ and $y_2(t) = e^{-t}$ are solutions. Further exploration leads to the fact that any function in the family $y(t) = c_1 y_1(t) + c_2 y_2(t)$, where c_1 and c_2 are arbitrary constants, is a solution. Computer algebra systems such as *Maple* are used to visualize the solution family for a range of c_1, c_2 values. As a next step the students explore and discover that specific solutions can be identified in the family by asking questions such as "Can we identify the solution which passes through the point $(0, 4)$ and has a slope -2 at that point?"

The explorations parallel theorems which demonstrate the steps needed to make the transition from computation to mathematical rigor. In what follows we assume the existence and uniqueness result for the initial value problem

$$L[y] = y'' + p(t)y' + q(t)y = 0 \quad \text{with} \quad y(t_0) = y_0 \quad \text{and} \quad y'(t_0) = y_0',$$

where $p(t)$ and $q(t)$ are continuous functions for $\alpha < t < \beta$. We do not delve into the details of this involved existence theorem. However, there are examples of theorems which parallel the explorations and use simple direct proof techniques. For example

Theorem 4 (Principle of Superposition) *If y_1 and y_2 are two solutions of the differential equation $L[y] = 0$, then the linear combination $c_1 y_1 + c_2 y_2$ is also a solution for any values of the constants c_1 and c_2.*

The question *can we find c_1 and c_2 so as to satisfy the initial conditions $y(t_0) = y_0$ and $y'(t_0) = y_0'$* leads naturally to the definition of a Wronskian and the next theorem.

Theorem 5 *Suppose that y_1 and y_2 are two solutions of $L[y] = 0$, and that the Wronskian $W(y_1, y_2) = y_1 y_2' - y_1' y_2$ is not zero at the point t_0 where the initial conditions are assigned. Then there is a choice of constants c_1, c_2 for which $y(t) = c_1 y_1(t) + c_2 y_2(t)$ satisfies the differential equation $L[y] = 0$ and the initial conditions $y(t_0) = y_0$ and $y'(t_0) = y_0'$.*

Finally, the students are led to the theorem characterizing the general solution.

Theorem 6 *If y_1 and y_2 are solutions of the differential equation $L[y] = 0$, and if there is a point t_0 where the Wronskian $W(y_1, y_2)(t_0)$ is nonzero, then the family of solutions $y = c_1 y_1 + c_2 y_2$ with arbitrary constants c_1 and c_2 includes every solution of $L[y] = 0$.*

Proving the last theorem requires more sophistication and maturity than the simple, direct proofs of the previous two. In particular, the students see how the uniqueness of solutions plays a fundamental part in the proof.

1.5.5 From Theory to Practice: Assessing the Outcomes of the Proofs Course

The authors received a grant (NSF Grant No. 0310753) from the National Science Foundation's Division of Undergraduate Education to develop the proofs class as part of the Adapt and Implement Track of the Course, Curriculum, and Laboratory Improvement (CCLI) Project (NSF-DUE-955414) and the Interdisciplinary Lively Applications Projects (NSF-DUE-9455980). It is organized in the spirit of Mount Holyoke's Laboratory in Mathematical Experimentation (NSF-DUE-9554646). Although designed to be a required course for sophomores, the course ran for the first time as an elective in Spring, 2004. Six of the 18 students enrolled in the class had taken upper level courses before enrolling in the proofs course, although none of them had taken our upper level courses in either algebra or analysis. These six students completed additional assignments to receive upper level elective credit, allowing the course to count for their major. The other twelve students had taken Calculus I and II and some were concurrently enrolled in Calculus III and/or Linear Algebra. These students received free elective credit and did not receive major credit.

We were able to team teach the course in Spring, 2004. Both Jones and Mukherjee were present for all classes minus the occasional missed class due to illness. The class met twice a week for 75 minutes over the 15 week semester. The class met in a room with a wireless hub, and we brought a computer cart of laptop computers into the classroom. Students accessed the Blackboard course management system using a wireless internet connection. The laptops had the latest versions of *Microsoft Excel* and *Maple*. Both of these software programs also were available in a public computer lab.

Students were assigned three group projects over the course of the term. These projects involved the mathematical modeling of problems involving things like mortgages, gasoline prices, and lizard populations. We adapted the technical report format of Miller's Technical Report Format and Writing Guide in [1] so that students were clear on the expectations for the written reports. Each project had an oral component where the students had to present their results. For the first two presentations, the teams (of two students) were given ten minutes to recount their results. Students were asked questions about their presentations by the instructors as well as by other students in the class. In general, we were pleased with the students' presentation skills, both written and oral. The oral presentation skills evolved significantly over the term and students were aware of their peers' abilities and techniques and would mimic successful styles. Even though we had suggested that the students summarize their results and expound on one aspect in detail, students recognized that successful presentations used less content on overhead or *PowerPoint* slides and did not try to present too much material.

We designed the first project to be straightforward so that students could focus more on the presentations than the mathematics of the project. While this approach was successful in that the students' first presentations were better than expected, we did miss an opportunity to highlight a more extensive application or ILAP. We spent time in class reading mathematics but we did not have an assignment that highlighted the importance of reading mathematics. Although we have used similar ideas in other classes, we did not motivate good writing through the reading of well-written mathematics. In Fall 2004 and Spring 2005, we required students to read a well-written mathematics article and answer questions on exposition before starting the first project.

During our first run through the course, we required students to turn in weekly homework assignments. We graded selected problems from these assignments and then discussed their solutions. We required students to write every homework problem well, although only certain homework problems required the students to write proofs. Although we did highlight the homework problems that required proofs, we did not divide the homework into two parts. We think that it is useful to specify which problems the students need to write up carefully. Subsequently we still assigned many homework problems, but specified which solutions were to be graded for their presentation. This allowed the students to concentrate more on their writing. To increase the oral communication skills of the students, we also required students to discuss the homework problems in class.

For the first two projects during spring 2004, it was possible for groups to emphasize the same mathematical content in their presentations. For the third or capstone project, each group had 15 minutes to give the details of a specific portion of the project. Each group was assigned specific content to explain, thereby eliminating any repetition in the project presentations. The final presentations were open for other faculty and students to attend. Besides the two faculty members from the class, an additional six faculty members attended the students' presentations of the capstone project. The student interactions during the capstone project presentations were ideal. Not only did students ask other groups questions during their presentations, the students also helped out their classmates when a group was unable to answer a question. We continued this practice during the 2004–2005 academic year.

We used *Microsoft Excel* and *Maple* worksheets to introduce concepts in the course. We projected the worksheets onto a screen at the front of the class and had the students follow along on their laptops. The worksheets and demonstrations showed the power and utility of the software programs and the projects required the students to use the programs to analyze the applied problems. All *Excel* files and *Maple* worksheets used in class were permanently stored on the Blackboard site for the course. These worksheets formed a reference library for the students to access when recalling the syntax of the *Maple* commands. Much of the student exploration was accomplished through their use of the software programs. We asked leading questions of the students and had them explore the consequences through *Excel* and *Maple*. Students were able to determine what would happen for ranges of parameter values by considering the outcome for different parameter values and then trying to understand why the outcomes occurred. This hands-on approach was useful in getting the students to make predictions and conjectures.

We introduced techniques of proof in 15 to 20 minute asides throughout the term. We assigned homework for the students to practice each proof technique. We introduced the proof techniques with content that had been introduced previously. The students thus were more comfortable with the content and were able to concentrate on the technique itself. For spring 2004, this was successful with direct proof and proof by induction, but we did not have the class time as the semester progressed to spend as much time on proof techniques as we would have liked. We altered this sporadic approach during the following terms by spending whole days on proof techniques early in each term.

The outline presented is ambitious. Implementing the syllabus was feasible, but we believe it is valuable to spend more time discussing proof techniques and continuing the exercises in exploration. The intangibles of exploration and discovery are more important than covering all of the aforementioned content for the course. We decided not to cover higher-order linear difference questions but instead to model higher-order linear difference equations as systems of first-order difference equations. Mark Parker, a mathematician from Carroll College and a co-organizer of the NSF-sponsored Project InterMath Workshop that motivated our proofs course served as an external evaluator of the grant. Some of the above suggestions for improving the course were given by Professor Parker as part of the evaluation process.

Along with the internal evaluator of the CCLI grant, Gideon Weinstein, a mathematics educator from Montclair State University, the authors helped design pre-class and post-class surveys to collect quantitative and qualitative data about the proofs course. Data from the surveys, as part of Weinstein's internal evaluation, and additional information on the course and its future iterations can be found at the website

http://www.csam.montclair.edu/~jonesma/transitions.html

1.5.6 Modifying this Course to Other Institutions

To directly apply our twist on a proofs course to another institution seems to require the conditions present at Montclair. In particular, the program should not offer a sophomore-level differential equations course and the program or department should be centered on applied mathematics. Of course, picking specific applications to match the research interests of the faculty depends on the faculty. The applications can be changed according to the program or the individual faculty. However, the transition course could be used *in lieu* of a sophomore level differential equations course in a department that also offers an upper level course in ordinary differential equations. The students then would receive some of the content taught in a typical sophomore-level differential equations course while gaining a broader perspective about what mathematics is about.

For institutions with more of a pure mathematics bent, the spirit of the proofs course can still be implemented with topics that coincide with faculty and program interests. This same benefit of having the students exposed to faculty research areas can increase the likelihood of students being ready to pursue research at an earlier point in their educational career. The exploratory aspect and use of technology can still be employed for a proofs course that surveys advanced pure mathematics. For example, there are software packages to demonstrate abstract algebra (e.g., the GAP software) and offer students an opportunity to explore and conjecture proofs.

1.5.7 Conclusion

The liberating idea of this article is that a proofs course can be tailored to fit the nuances of a specific program and does not have to be a survey of pure mathematics or an introduction to logic. Teaching students about how to read and write proofs can be accomplished regardless of the mathematical content. As opposed to other more advanced courses, proofs courses are less motivated by content and more concerned with providing an environment where students can explore and make conjectures and see how mathematics is done as opposed to how it is learned. We chose to highlight applied mathematics to coincide with faculty interests and the realization that few of our students go on to graduate school. If the population of students at Montclair changes, our proofs course is easily adaptable to such changes.

Note: This material is based upon work supported by the National Science Foundation under Grant No. 0310753. Any opinions, findings, and conclusions or recommendations expressed in this material are those of the authors and do not necessarily reflect the views of the National Science Foundation.

References

1. D.C. Arney, *Interdisciplinary Lively Applications Projects (ILAPs)*, Mathematical Association of America, Washington, DC, 1997.
2. W.E. Boyce and R.C. DiPrima, *Elementary Differential Equations and Boundary Value Problems*, 6th Edition, John Wiley & Sons, Inc., New York, NY, 1997.
3. P.J. Campbell, *UMAP/ILAP Modules 2000: Tools for Teaching*, COMAP, 2000.
4. M. Eisenberg, *The Mathematical Method: A Transition to Advanced Mathematics*, Prentice Hall, Upper Saddle River, NJ, 1996.
5. P. Fletcher and C.W. Patty, *Foundations of Higher Mathematics*, 3rd edition, PWS-KENT Publishing Company, Boston, MA, 1995.
6. D.L. Isaacson and R.W. Madsen, *Markov Chains: Theory and Applications*, Robert E. Krieger Publishing Company, Inc., Malabar, FL, 1985.
7. "It's only a game, *The Economist*, April 13, 1996.
8. R. Larson, R.P. Hostetler, and B.H. Edwards, *Calculus*, 7th Edition, Houghton Mifflin Company, Boston, MA, 2002.
9. Mount Holyoke College, *Laboratories in Mathematical Experimentation*, Springer-Verlag, New York, NY, 1997.
10. J.J. Rotman, *A Journey into Mathematics: An Introduction to Proofs*, Prentice Hall, Upper Saddle River, NJ, 1997.
11. C. Schumacher, *Chapter Zero: Fundamental Notion of Abstract Mathematics*, 2nd Edition, Addison-Wesley, Boston, MA, 2001.
12. B. Sinervo, `http://www.biology.ucsc.edu/~barrylab/#lizardland`, 2003.
13. B. Sinervo and C.M. Lively, "The rock-scissors-paper game and the evolution of alternative male strategies, *Nature*, 340 (1996) 240–246.
14. D. Solow, *How to Read and Do Proofs: An Introduction to Mathematical Thought Processes*, 3rd Edition, Wiley & Sons, Inc., New York, NY, 2002.
15. D.J. Velleman, *How to Prove It: A Structured Approach*, Cambridge University Press, Cambridge, MA, 1994.
16. D.B. West and J.P. D'Angelo, *Mathematical Thinking: Problem-Solving and Proofs*, 2nd Edition, Prentice Hall, Upper Saddle River, NJ, 2001.

17. J.W. Weibull, *Evolutionary Game Theory,* MIT Press, Cambridge, MA, 1995.

Brief Biographical Sketches

Michael A. Jones received his PhD in Game Theory from the Department of Mathematics at Northwestern University in 1994. After a three-year temporary position at the United States Military Academy at West Point and a year visiting Loyola University in Chicago, he has been at Montclair State University in Upper Montclair, NJ in a tenure track position. While busily worrying about how students can become more aware of their opportunities while transitioning to higher mathematics, he himself transitioned by receiving tenure in Fall 2003.

Arup Mukherjee received his PhD in Applied Mathematics from the Department of Mathematics at Penn State in 1996. After spending four years at Rutgers University in New Jersey, he is currently an Assistant Professor at Montclair State University in Upper Montclair, NJ. In addition to worrying about his students and their mathematical maturity, he constantly strives to improve on his maturity, as well as his knowledge of computational methods for partial differential equations.

Chapter 2

Course-Specific Papers

Introduction 55

2.1 Wrestling with Finite Groups; Abstract Algebra need not be Passive Sport, *Jason Douma* 57

2.2 Making the Epsilons Matter, *Stephen Abbott* 67

2.3 Innovative Possibilities for Undergraduate Topology, *Samuel Bruce Smith* 81

2.4 A Project-Based Geometry Course, *Jeff Connor and Barbara Grover* 89

2.5 Discovering Abstract Algebra: A Constructivist Approach to Module Theory, *Jill Dietz* 101

Introduction

The second chapter contains five papers that describe approaches to core courses in the undergraduate major that excite student interest while delivering solid mathematics courses. In the first paper, Jason Douma of the University of Sioux Falls discusses how an abstract algebra course can be organized around an open-ended research project. The project is not an application of material presented in class but rather serves to motivate and generate the course content. In the same vein, Jill Dietz of St. Olaf's College used a guided discovery approach to generate student input and ideas that eventually lead to a course in module theory as a follow-up to an introductory course in abstract algebra. In both cases, students are expected to be extremely active and, with appropriate guidance, develop the course material on their own. Both papers contain a good deal of supplementary material to support implementation of the respective approaches.

This theme continues in the geometry article by Jeff Connor and Barbara Grover of Ohio University. In this case however, the students are expected to generate axiom systems for both Euclidean and non-Euclidean geometries, using technological supplements when appropriate. Likewise, Samuel Smith of St. Joseph's University works to maximize student participation in developing a topology course that is intended to appeal across the board and not just to students planning to do graduate work. The key in this case is using an initial geometric approach to motivate the axiom structure that characterizes topology.

Students in an introductory course in real analysis must develop a solid understanding of continuity, differentiability, integrability, and convergence in order to probe more deeply into the world of analysis. At the same time, the course should not look like calculus with epsilon and delta proofs added. Stephen Abbott of Middlebury College shows how this can be accomplished through a series of narrative tutorials that are done by students working in groups. Once again there is an emphasis on student engagement in the learning process.

The five papers in Chapter 2, while more course specific than their counterparts in Chapter 1, still contain approaches that are adaptable in a variety of settings. They also continue the theme of encouraging students to move beyond the bounds of the standard course content. They provide a useful contrast to the papers in Chapter 3, which invoke concepts, materials, and methods that were not typical just a few years ago.

2.1

Wrestling with Finite Groups; Abstract Algebra need not be Passive Sport

Jason Douma
University of Sioux Falls

2.1.1 Introduction

Abstract algebra *is* fundamentally abstract. (Perhaps this is stating the obvious.) But must the teaching and learning of abstract algebra take place in an entirely passive environment? Certainly not.

In the 2001 spring semester at the University of Sioux Falls, I structured an abstract algebra course around a single open-ended research project. The project was not an application or assessment of material already covered in class. This central project was intended to motivate and introduce much of the actual course content throughout the entire semester.

The class project called for students to catalog the central product structures of all finite 2-groups through order 32, and to articulate and prove a general statement about the central product structures of all abelian 2-groups. Because the class was small (six students), all of the students were expected to carry out the project as a single collaborative group.

The experiment appears to have been quite successful and could transfer well to other advanced mathematics courses, especially if the class size is small. I suggest that there are at least three key ingredients that should be present in the design and administration of your own project-driven course: 1) the semester project should be sufficiently open-ended to allow the students an authentic research and discovery experience; 2) the project topic should be rich enough to trigger many of the concepts that are essential to the course's syllabus; and 3) the instructor should be prepared to be flexible and responsive; reading word for word from a detailed set of lecture notes would defeat the very purpose of a project-driven course.

2.1.2 Objective

For many undergraduate mathematics students, a semester course in abstract algebra is perhaps their first and most authentic taste of pure mathematics. Here students see some truly beautiful structures and powerful theorems. In many ways the beauty, elegance, and purity of the objects we study in abstract algebra must be credited to their essentially abstract nature. Indeed, it was in my first semester of abstract algebra as an undergraduate mathematics student that I first began to *love* mathematics, just as a poet loves poetry or a musician loves music. Until that point, I saw myself as more of a mathematical technician: capable and effective, but with no particular aesthetic attachment to the objects of my study.

However, the same abstractness that is capable of delivering such sublime elegance also seems to foster a noticeable passivity in learning abstract algebra (and in other upper-division pure mathematics subjects). All too often, the student's approach is to receive information from the teacher and attempt to intellectually store this information *in exactly the same form as its original presentation by the teacher*. They treat the content as if it were too brittle to withstand actual handling or use. Nadine Myers has provided an apt description of this pattern [5]:

> Many students of mathematics find abstract algebra to be the most difficult undergraduate course they encounter... Even good students find themselves floundering in abstraction, struggling to prove statements about mathematical objects that are themselves elusive and only superficially understood.

To the extent that this form of passive and superficial learning is present, the student is left with a less than authentic understanding of the mathematics and a far less than authentic experience in the creative process that generates and sustains mathematics.

This article outlines one effort to encourage abstract algebra students to come into meaningful contact with their objects of study, to handle these structures, and even to wrestle with them during the course of a semester.

2.1.3 Background

The University of Sioux Falls is a small private liberal arts college affiliated with the American Baptist Churches. Enrollment has grown to about 1400 students, just under 1000 of whom are full-time undergraduates (we do have graduate programs in education and business). The two largest majors on campus are in professional fields (business and education), but the university does maintain a healthy core of students majoring in one or more of the arts and sciences.

Mathematics majors at the University of Sioux Falls form a close-knit group and students typically complete all of their upper-division courses with roughly the same band of classmates. Enrollment in upper division mathematics courses is modest but stable, generally between 5 and 12. Students are comfortable working together both in and out of class and justifiably expect continued interaction with faculty and classmates beyond the class meeting times.

Prior to the 2001 spring semester, our abstract algebra course consisted of the usual survey of essential topics in abstract algebra. Roughly half of the semester would concentrate on group theory with the balance devoted to rings and fields, including the associated topics concerning number theory and divisibility. A second semester of abstract algebra is not offered on a regular rotation at USF (only as a special topics course or directed study). Consequently, this traditional sampling of topics has been the extent of most of our students' contact with abstract algebra.

The abstract algebra course takes on some of the characteristics of a capstone course at the University of Sioux Falls. It draws on some of the content background from other courses (especially linear algebra) and builds strongly on the modes of reasoning (proofs) developed in other courses. A significant research project is typically included in the syllabus. Indeed, students often use their work in abstract algebra as a springboard for the work they present in our Natural Science Colloquium.

2.1.4 Searching for a New Approach

I began the planning process for this course with several familiar concerns and fine opportunities firmly in mind. I wanted to find a way to counteract the gravitational pull of passive learning in this upper-division course. I also wanted to provide my students with a meaningful, open-ended project that would give them a more authentic experience with the creative process through which mathematics is developed and refined. Ideally, such a project might lead to results that students could present (or publish?) in a more substantial forum.

As acknowledged earlier, I also had some key assets at my disposal: a small class size and a collection of students who were comfortable working together. Meanwhile I had also been culturing some questions and curiosities concerning central product structures of groups. These questions had surfaced during the course of my doctoral work but I had not found time or occasion to revisit them. Since central products essentially are little more than a generalization of the direct product, it seemed that many of my questions might be accessible to my undergraduate students. Here then was a collection of material that could serve as the "meaningful, open-ended project" at the heart of the abstract algebra course.

The new course paradigm had taken its initial shape. The course would be built around a single, common, open-ended project related to the central product structures of finite groups. (The precise wording of the research problem is given in a later section.) Students would pursue the project collaboratively and in the process they would learn about group structures by actually handling and dissecting these objects, rather than merely hearing or reading about them.

Of course, this approach raised some concerns that would be familiar to anyone who has sought to augment the level of interaction in their mathematics classes. In particular, the project would require a substantial investment of precious class time. Furthermore, the project would have to be up and running *before* the prerequisite knowledge is in place. A project of this scope could scarcely be addressed in a single semester. If we were to wait until the fifth or sixth week to introduce the project, we might surrender any reasonable hope of reaching even a tentative conclusion by semester's end. Clearly, this project could not be a supplement to material already covered in class. There simply would not be enough time. Rather, in some sense, the project would need to *be* the content of the course, even in the first weeks of the semester.

Many of these changes would not come easily for me. I have always found it easier to add topics to the course syllabus than to delete them. After all, there are always so many interesting topics to discuss (especially in abstract algebra) and I cringed at the thought that my students might miss out on something. I had also had mixed success with collaborative work in my other classes. In some cases, students did seem to achieve a higher degree of ownership of their learning, but often it seemed as though their wanderings were inefficient at best and aimless at worst. Nevertheless, the potential benefits of the new course structure appeared to outweigh the apparent risks and sacrifices. I resolved to pare back my beloved syllabus and trust more of the content development to the students themselves.

2.1.5 Allocating Time for the Project

To allow for appropriate attention to the project as an authentic vehicle for developing content, the syllabus was designed so that class time could be apportioned into three equal parts.

1/3 scheduled lecture (independent of the project)

1/3 workshop time, when students would attend to the project together during class time

1/3 discussion and lecture motivated by the project

(Retrospective note: In practice, I deviated a bit from this uniform distribution. The students quickly demonstrated the ability to work productively together outside of class, and thus the actual distribution of class time settled into something closer to (2/5, 1/5, 2/5) among the three categories.) Implicit in this curricular plan was the assumption that the demands of the project itself would be able to motivate and drive a great deal of the learning that was to take place in the coming semester.

As a concession to the demands of the project, coverage of rings and fields was scaled back to approximately three weeks at the end of the semester. Students would still learn the terminology, key examples, and important results from ring theory. However, most results would be presented without proof, and the overall degree of rigor would be lower than that which was given to group theory. I viewed this downsizing of ring theory as necessary, but not at all pleasing. Ring theory is every bit as important and interesting as group theory. Many fine instructors and texts (e.g., Hungerford [4]) have placed ring theory at the beginning of their abstract algebra course. Nonetheless, a substantial investment of time would be required for proper development of the project, which happened to be in the area of group theory.

2.1.6 The Project Assignment

The essential idea that eventually evolved into the abstract algebra class project can be traced back a few years to some of the work related to my dissertation [1]. While investigating automorphisms of direct products of finite groups, I found that central product structures would play a key role in determining the nature of the maps in question. A literature search revealed that even though the basic principles associated with central products are simple and well understood, there is a dearth of published reference material on the subject. Surely other mathematicians have thought about central products in the past, perhaps as required in their own research. Nonetheless, little effort appears to have been made to document or compile what is known about them. No equivalent of Thomas and Wood's *Group Tables* [6] appears to exist for central product structures of groups.

Like others who had gone before me, I developed just enough understanding of central products to address the research problem at hand. I did, however, make a mental note that there was more to be learned in this area. In particular, this appeared to be fertile ground for undergraduate research. There were few prerequisites (just basic group theory) and questions from this area were open-ended in the sense that there was apparently no central, accessible source of answers to such questions.

The task of developing a research question for the abstract algebra class project had now been reduced to specifying a set of directives that was modest enough to be addressed within one semester and yet comprehensive enough to motivate much of the introductory group theory curriculum. The following research problems appeared to accomplish a reasonable balance, and were presented to the class as their common project for the semester.

Catalog and classify the central product structure of all finite 2-groups through order 32.

Seek and prove a more general result concerning central product decompositions of abelian 2-groups.

For the sake of the reader with some interest or background in central products, I should clarify that I adopted Gorenstein's characterization of the central product [3], which allows the overlap between subgroup factors to be any subgroup of the group's center and not necessarily the entire center of the group. This permits the central product to be viewed as a generalization of the direct product, where the overlap between subgroup factors would simply be the trivial subgroup. The following definition serves to clarify the distinction, since there is some ambiguity in the existing literature.

Definition. A finite group G is a *central product* of normal subgroups

$$G_1, G_2, \ldots, G_n \text{ if } G = G_1 G_2 \cdots G_n \text{ and } [G_i, G_j] = 1 \quad \forall \, i \neq j.$$

(Note that if $x \in G_i \cap (G_1 G_2 \cdots \widehat{G_i} \cdots G_n)$ for some i, then the definition implies that x is in the center of G.)

What we are saying here is that a group G is a central product of two of its normal subgroups H and K if $G = HK$ as an internal product and $[H, K] = 1$. Note that this definition implies that any elements H and K have in common must be contained in the center of G. Thus an internal direct product is really just a special case of the central product. (In this statement, $[H, K]$ refers to the commutator subgroup consisting of all commutators of h and k, where h is an element of H and k is an element of K.)

2.1.7 The Project as Source of Content

These research problems were unveiled on the first day of class with the understanding, of course, that none of the directions would be intelligible at the very onset. The initial groundwork was laid in the first class session by comparing the task at hand to other more familiar product structures that may be used to organize and classify mathematical objects: factoring polynomials, prime factorizations via the fundamental theorem of arithmetic, etc. With this basic motivation in place, students' attention naturally turned to understanding *what sort* of product structures are being specified through the terminology in the research problem. They needed to know what a finite 2-group is, what is meant by order, and certainly what a central product entails. Each of these concepts, in turn, rests on other notions and terms that must be developed in sequence. Thus, within the first two class sessions, the basic terminology and definitions associated with group theory were introduced into discussion, just as they would be in a traditional abstract algebra course, with one notable difference: the students were already viewing the material with an eye to *doing something with it*. The demands of the project had motivated the development of the course content.

As the semester progressed, I attempted to keep a running list of the group-theoretic content motivated by the project. By the end of the course, the list was fairly impressive.

- definition of *group*, *subgroup*, and *element*
- the concept of *order* for groups, subgroups, and elements
- examples of finite groups
- the use of group tables and subgroup lattices
- center
- fundamental theorem of cyclic groups
- normal subgroups
- internal products and central products
- cosets and factor groups
- direct products
- commutators and abelianizations
- Lagrange's theorem
- p-groups
- homomorphisms and isomorphisms
- kernel

- first isomorphism theorem
- automorphisms and automorphism groups
- fundamental theorem of abelian groups
- generators and relations

This list contains most of the traditional group theory syllabus from a typical first-semester abstract algebra course. Among the notable exceptions are permutation groups, alternating groups, Cayley's theorem, and the orbit-stabilizer theorem. (Not surprisingly, permutation groups and related topics do not arise as naturally in a project that is focused on finite 2-groups.) These important topics from group theory along with all of the course's content from ring theory were introduced in a more traditional manner, independent of the class project. We used Gallian [2] as our principal text, both for reference in many of the matters related to the class project and for development of concepts outside the scope of the project. Thomas and Wood [6] provided an essential reference for the detailed structure of each of the groups being studied. Along the way, students also referred to a few tables and to results from my dissertation.

Much to my delight, there were days when the project motivating content paradigm worked to perfection. I recall one particular day when I walked into the classroom and was greeted by the question, "So, are direct product decompositions of groups *unique*?" That students would have the insight to ask this question on their own accord is pleasing enough, but what impressed me most of all was the sight of these students *genuinely interested* in what the answer to this question might be. They were hanging on my every word (and believe me, it was not due to anything extraordinary in my lecturing style). There were other days when the class would update me on the status of their project, and their statements would flow seamlessly into a discussion of some key concept from group theory.

Still, there were also many days when the students had trouble knowing which question to ask. They were unable to anticipate what they needed next or could not identify the void that had inhibited their progress. On these days, the project still motivated content in the sense that I was able to introduce material that would help them on their way, but the exchange of information was somewhat more of a one-way street.

2.1.8 Creative Output

There is a sense of pride and reward to be found for both the student and the teacher when the results of a semester's toil amount to more than just numerical scores on assignments (and exams) that were concocted as little more than assessment tools. We all long to *produce* something meaningful. Through their study and labor, the students in this abstract algebra class were able to create something new and meaningful in its own right.

Evidence of students' pride in their creative work can be found even in the names given to some of the methodologies they developed.

The Birger Method (for ascertaining the orders of individual group elements when expressed as products of elements from central product factors)

Mike's Six-Step Program (for checking whether a given external product of smaller groups is isomorphic to a larger specified group)

It was clear from both the nomenclature and student attitudes that the class had taken ownership of their work. Rather than following a preordained script or responding to exercises (as in an exam) that had been handed to them, these students had produced something creative and meaningful from their own efforts.

Because their work had meaning and value beyond the context of the abstract algebra course itself, the students were given opportunities to present their findings to broader audiences. Several USF undergraduates attended the 2001 Math on the Northern Plains Conference at the University of South Dakota in April, where one student presented a preliminary report on behalf of the class. For a few of the students, including the presenter, this was a first experience with conference talks and proceedings. Later in the semester, the class shared their results publicly with members of the campus community. Two students from the class have also further developed ideas from the class project for presentation in the USF Natural Science Colloquium. (All USF students majoring in any of the natural sciences are required to give two research talks at the colloquium.)

Of course, the centerpiece of the students' creative output consisted of their actual findings in response to the research problems. The class produced a collection of tables that listed and classified the central product structures of each of the finite 2-groups of order 2^k, where $1 \leq k \leq 5$. Many of these structures were indeed trivial; but even at these relatively low orders, there are a number of groups with subtle central product structure. For example, there exists a group of order 32 (group 32/17, following Thomas and Wood's indexing [6]) that can be expressed as three distinct central products: $\Gamma_2 b\ \mathbf{Z}_8$, $\Gamma_2 d\ \mathbf{Z}_8$, and $D_4 \mathbf{Z}_8$. In the first two cases, a copy of $\mathbf{Z}_4$, a proper subgroup of the center of the group, is shared between the two factors. In the third case, the common elements make up a copy of $\mathbf{Z}_2$ from within the center of the group. Clearly there is no direct equivalent of the Krull-Schmidt theorem for central product structures. The students, through their active handling of these objects, had an immediate understanding of this fact and the subtle consequences surrounding it. The class did find, in response to the second research problem, that there *is* something similar to the Krull-Schmidt theorem in the case of abelian 2-groups. This was presented as a theorem with proof in their final paper.

2.1.9 Assessing Student Development

The collaborative research project served not only as the focal point for student activity but also as the focal point for assessment. Student development relative to the project was assessed through four vehicles: a progress report completed by each student near the middle of the semester, the research paper, the public research presentation, and a personal reflection submitted by each student at the end of the semester.

The progress report served as a device for evaluating individuals within the collaborative effort. Each individual received a separate score on the progress report, based on their contribution to and understanding of the project up to that point in the semester. This also serves as a mechanism for accountability and early intervention if there were unacceptable disparities among individual contributions. To complete the progress report, each student submitted a brief summary of the status of the research project, as well as an evaluation of every member of the class (including themselves) in the four areas of reliability/attitude, project leadership, content expertise, and product production. Table 2.1.1 shows the rubric used for the evaluations. The header row indicates the points awarded for each rating. A student's score in each category consisted of the sum of the individual ratings he or she received in that category, with the lowest and highest ratings omitted.

All students in the class received a common score for the research paper itself. The public presentation was graded in a manner that allowed for a combination of individual and common scoring. The personal reflection paper was, of course, an individual effort.

In aggregate, the assessment related to the class project accounted for 40% of the semester grade. A brief descriptive essay, take-home exams, and a one-on-one oral final examination constituted the balance of the student assessment.

Category	0 (poor)	2 (marginal)	4 (acceptable)	5 (superior)
Reliability & Attitude	missed several meetings or deadlines; doesn't seem to want to contribute	contributes occasionally, but is unreliable or resentful of responsibility	generally contributes willingly, but has missed one or two meetings with good reason	contributes positively without exception
Project Leadership	only acts when directly asked to do so; seems unaware of project goals	understands project goals, but is usually unable to identify the "next step"	is able to articulate the direction of the project with assistance from the group	astutely assesses project status and provides direction for next steps
Content Expertise	unable to discern correct conclusions from false conclusions	can verify the work of the group, but rarely produces results independently	produces accurate results when called upon, sometimes with assistance	identifies generalizations and structure and provides reliable computations
Product Production	maintains only inaccurate or incomplete records of the group's work	keeps personal copy current, but adds little to production of the project document	valuable word processor or data processor	provides vision for concept and composition of the project document or database

Table 2.1.1. Progress Report Rubric

2.1.10 Assessing the Course

The oral final examinations provided a source of comparative course assessment between the project-centered course offered in the 2001 spring semester and a more traditionally formatted section of the same course I had taught two years earlier. Both classes made use of an oral final examination with a similar set of questions and prompts. (There was just one significant change in content: the exam from the traditional course contained four questions from ring theory; the exam from the project-centered course featured two questions from ring theory and two questions on the theory of central products. The remaining questions from group theory were similar, if not identical, in both exams.) Both exams used the same rubric for assigning points to the student's response. Both exams also offered up to four bonus points along the way, with a total of 50 points possible.

While the possibility of some subjectivity in grading must be acknowledged, scores did improve from the traditional section ($n = 8$, $\bar{x} = 42.5$, $s = 6.4$) to the project-centered section ($n = 5$, $\bar{x} = 46.0$, $s = 3.0$), although not significantly ($p = 0.281$), due in large part to the modest sample sizes. Qualitatively, the students from the project-centered class appeared far more comfortable discussing these topics spontaneously than their traditional class counterparts.

Student evaluations of the course were also positive, as reflected in comments such as these:

> I really liked the way the lectures in class gave us the tools we needed in the project when we needed them so we could see their application right away.

> We accomplished much more together than we would have been able to alone. I think that this is a great model to follow in the future.

Many other student comments were variants of the theme, "It was a lot of work, but I understand the material more deeply as a result."

Most students did express concern that at times they felt they were working double shifts; continuing work on the class project while also working just as hard to keep up with the textbook part of the course. As is often the case with collaborative work, the students also felt exasperated in their efforts to coordinate schedules and find times to meet. These are legitimate concerns that perhaps could be addressed more through adapting and clarifying expectations than through major changes in the course content and structure.

In planning a course in which a central project would motivate much of the course content, I had hoped, as one of the learning objectives of the course, that students would further develop their ability to identify and articulate key questions that are pivotal to success in their work. By midterm, the group had begun to ask such questions with greater frequency and sophistication than students in previous upper-division courses I had taught. But as the semester progressed, it seemed that two or three students emerged as spokespersons for the class. The remainder of the group was arguably as tacit in this context as they would have been in a more traditional setting.

On the whole, it appears that student learning in the project-centered course was more active, deeper (though perhaps a bit narrower), and more fulfilling than in a traditional abstract algebra course. I do intend to continue this approach in future abstract algebra courses.

2.1.11 Conclusion

The instructor who designs and leads a project-driven course is taking a risk by breaking with traditional (and comfortable) notions of control and coverage. However, the potential for authentic student engagement is great enough to make this a risk worth venturing, subject to the availability of a few key factors.

Without question, the most critical element in the design of a project-centered course is the choice of research problem. The instructor is hunting for a gem. The effective research problem must be modest enough to be addressed within a single semester, comprehensive enough to motivate a substantial share of the course content, and open-ended enough to provoke genuine student discovery. (Do note that open-ended need not imply that the project topic is a recognized *open question* in the field.) Surely these topics exist in crevices and under rocks throughout the mathematical landscape, not just in abstract algebra. The instructor is perhaps most apt to recognize these gems in areas near her neighborhood of expertise. A project-centered approach demands that the instructor anticipate how the project is likely to unfold and to react wisely to unexpected developments. For these reasons, the project-centered approach is likely most ideal for upper- division courses near the instructor's area of expertise. For the same reasons, I do not feel qualified to recommend possible project topics for courses other than abstract algebra, although I am confident that they can be found with the help of the instructor's insight, creativity, and expertise.

Class size is also an important factor. Especially at a residential college, groups of up to six students might reasonably be expected to work together in a major endeavor such as this. For slightly larger classes, it might be feasible to have two or three groups working in parallel, perhaps on different features of the same larger project. Periodically, the groups could share their findings and compare notes. Because the course structure relies on the project to provide a substantial portion of the content, it may not be appropriate to assign substantively different projects to each group in a class with multiple groups. Instructors with much larger classes might find that the individual attention required from the instructor cannot be adequately distributed among many groups.

Teaching the project-centered course is an exercise in responsiveness. The instructor is in many ways more of an interpreter than a pilot, since the students will strongly influence the development of the course. The result is stimulating and quite fruitful.

References

1. J. Douma, *Automorphisms of Products of Finite p-Groups with Applications to Algebraic Topology* [dissertation], Northwestern University, Evanston, Illinois, 1998.

2. J. Gallian, *Contemporary Abstract Algebra*, 4th edition, Houghton Mifflin, Boston, 1998.
3. D. Gorenstein, *Finite Groups*, Chelsea, New York, 1968.
4. T. Hungerford, *Abstract Algebra: An Introduction*, Saunders College Publishing, New York, 1997.
5. N. Myers, "An oral-intensive abstract algebra course," *PRIMUS*, 10 (2000)193–205.
6. A.D. Thomas and G.V. Wood, *Group Tables*, Shiva Mathematics Series 2, Shiva Publishing, Orpington, Kent, 1980.

Brief Biographical Sketch

Jason Douma received his BA from Gustavus Adolphus College and his PhD in mathematics from Northwestern University. He is now an Associate Professor of Mathematics and director of the liberal arts honors program at the University of Sioux Falls. His mathematical interests include group theory and the philosophy of mathematics.

2.2
Making the Epsilons Matter

Stephen Abbott
Middlebury College

Our subject is the most curious of all—there is none in which truth plays such odd pranks.
—G.H. Hardy

2.2.1 Calculus or Analysis?

My first attempt at teaching an introductory course in real analysis went well enough I thought. The students came to understand the logical structure of the proper definition of a limit and we used it to prove that polynomials really are continuous. I introduced enough topology of the real line to show that continuous functions on compact sets are uniformly continuous and attain extreme values, and then pressed on to show how this leads to an elegant proof of the Mean Value Theorem for the derivative. In the last part of the term we made a proper pass through the theory of the Riemann integral and, as a big finish, used our rigorously justified Mean Value Theorem to construct an argument for the Fundamental Theorem of Calculus. When the dust settled there was plenty to be proud of. The course evaluations were generally positive, the students learned how to write a proper ϵ–δ proof and, as far as I could tell, no one had gotten hurt along the way.

Although it took several years of thinking and tinkering before I was able to put my finger squarely on why my first versions of this course felt oddly unsatisfying, the conclusions I reached are hardly revolutionary. If I had asked the best students from that first semester to characterize real analysis, they would have explained that it is a careful retracing of the introductory calculus syllabus where we take the time to fill in the gaps and flesh out the thorny details of the proofs. The problem with this answer is that, although it describes the class I taught, it really does not describe the subject. To the best of my knowledge, the continuity of polynomials has never been in any doubt, and although a proof of this fact may be good evidence that our definition of continuity is reasonable, it is certainly not the reason analysis was created.

For me, the missing ingredient in my course was a worthy reward for the the hard work of firming up the logical structure of limits. A proof of the Chain Rule, or even of the Fundamental Theorem of Calculus for continuous functions does not in my opinion reach the level of reward status. These are familiar places—too familiar in the sense that students (like the pre-19th century founders of calculus) have already enjoyed considerable intuitive success with these topics *without* the tools of analysis. The goal of a course in analysis should be to challenge and improve mathematical intuition, not verify it.

Truth plays such odd pranks

The painstaking process of learning to read, write and communicate in the language of rigorous proofs makes a required course in analysis a widely dreaded event for students of mathematics from institution to institution. With 10 or 15 years of calculus reform efforts emphasizing graphical and numerical ways of thinking, it is tempting to suppose that we are doing our students a favor by limiting the scope of introductory analysis to the familiar terrain of first year calculus. The problem is that by trying to make the course easier we inevitably make it less interesting and consequently less worthy of the effort it requires.

A much better idea is to trust in the curiosity of our students and the beauty of our subject. A first course in analysis is often seen as a preparatory course for developing the skills to go on to investigate the enigmatic world of "odd pranks" associated with functions of a real variable, but I would argue that it should be the place where these questions are confronted head on. Now it is certainly the case that most textbooks, and most courses, already do this to some degree. To motivate completeness we generally start by showing that not all numbers are rational. Before discussing cardinality we might wonder whether there are more rational numbers or irrational numbers. When distinguishing between absolute and conditional convergence it is natural to investigate whether the terms in an infinite sum can be rearranged without affecting the value or the convergence of the series. The surprising answers to these questions give the course its electricity and ultimately make the axioms and epsilons matter.

The point I wish to make is that even in a one-semester course—in fact, especially in a one semester course—we should carry this motivational philosophy through to all of the topics. Does the Cantor set contain any irrational numbers? Can the set of points where a function is not continuous be arbitrary? Are derivatives continuous? If not, can any function be a derivative? Is every continuous function a derivative? Is every continuous function differentiable somewhere in its domain? If not, are most continuous functions differentiable at some points? Is every infinitely differentiable function the limit of its Taylor series? Using the Riemann integral, is it possible to integrate every derivative? If not, is there some way to modify the Riemann integral so that we can?

By shifting the emphasis to topics where an untrained intuition is severely disadvantaged, the hard work of a rigorous study of functions becomes a much more reasonable request. Students wandering through uncharted intellectual terrain are much more likely to buy into an axiomatic approach when they are investigating questions that are inaccessible without it.

A shift of emphasis, not content

Over the years, I have made mental notes every time I hear myself say to the class, "Well, it turns out that..." My goal in each case then becomes to carve out a rigorous path to these results using only the standard ingredients of a traditional first semester course in analysis. (For me, these topics consist of completeness and compactness in **R**, sequential and functional limits, continuity, uniform convergence of sequences and series, differentiation and Riemann integration.) What is interesting is that, over time, the raw material of the course has not changed much from year to year. What has changed a great deal is where I place the emphasis. We spend less time verifying that we can evaluate certain integrals with anti-derivatives and more time investigating what we can integrate. We spend less time proving power series are well-behaved and more time investigating why Fourier series are not.

In most cases, I have taken these challenging topics and worked them into narrative tutorials that I assign as group projects for two or three students. A corollary to this procedure is that, although at first glance it would seem that adding some advanced topics might overwhelm the weaker students, the level of difficulty can still be adjusted a great deal by altering the parameters and the frequency of the assignments.

2.2.2 Sample Assignments

What follows are some examples of the projects that I have created. The time period for this kind of assignment is usually a week and a half, and I make sure I am available for regular consultations. More than once I have set assignments like these in place of a final exam and used our designated exam slot as a time for student presentations (and heavy snacking).

Sets of Discontinuity

Given a function $f : \mathbf{R} \to \mathbf{R}$, define $D_f \subseteq \mathbf{R}$ to be the set of points where the function f fails to be continuous. We have seen that Dirichlet's nowhere continuous function $g(x)$ has $D_g = \mathbf{R}$, and a modification $h(x)$ of Dirichlet's function has $D_h = \mathbf{R}\setminus\{0\}$, zero being the only point of continuity. Finally, for Thomae's function $t(x)$, we showed that $D_t = \mathbf{Q}$.

Exercise 1 Using modifications of these functions, construct a function $f : \mathbf{R} \to \mathbf{R}$ so that
   (a) $D_f = \mathbf{Z}$.
   (b) $D_f = \{x : 0 < x \leq 1\}$.

The question to be investigated here is whether D_f can take the form of *any* arbitrary subset of the real line. In fact, we shall prove that this is *not* the case. The set of discontinuities of a real-valued function on $\mathbf{R}$ has a specific topological structure that is not possessed by every subset of $\mathbf{R}$. Specifically, D_f, no matter how f is chosen, can always be written as the countable union of closed sets. In the case where f is *monotone*, these closed sets can be taken to be single points.

Monotone Functions

Classifying D_f for an arbitrary f is somewhat involved, so it is interesting that describing D_f is fairly straightforward for the class of monotone functions.

Definition 1 A function $f : A \to \mathbf{R}$ is *increasing* on A if $f(x) \leq f(y)$ whenever $x < y$ and *decreasing* if $f(x) \geq f(y)$ whenever $x < y$ in A. A *monotone* function is one that is either increasing or decreasing.

Continuity of f at a point c means that $\lim_{x \to c} f(x) = f(c)$. One particular way for a discontinuity to occur is if the limit from the right at c is different from the limit from the left at c. As always with new terminology, we need to be precise about what we mean by "from the left" and "from the right."

Definition 2 (Right-hand limit) Given a limit point c of a set A and a function $f : A \to \mathbf{R}$, we write

$$\lim_{x \to c^+} f(x) = L$$

if for all $\epsilon > 0$ there exists a $\delta > 0$ such that $|f(x) - L| < \epsilon$ whenever $0 < x - c < \delta$.

Equivalently, in terms of sequences, $\lim_{x \to c^+} f(x) = L$ if $\lim f(x_n) = L$ for all sequences (x_n) satisfying $x_n > c$ and $\lim(x_n) = c$.

Exercise 2 State a similar definition for the left-hand limit

$$\lim_{x \to c^-} f(x) = L.$$

Theorem 1 *Given $f : A \to \mathbf{R}$ and a limit point c of A, $\lim_{x \to c} f(x) = L$ if and only if*

$$\lim_{x \to c^-} f(x) = L \quad \text{and} \quad \lim_{x \to c^+} f(x) = L.$$

Exercise 3 Supply a proof for this proposition.

Generally speaking, discontinuities can be divided into three categories:

(i) If $\lim_{x \to c} f(x)$ exists but has a value different from $f(c)$, the discontinuity at c is called *removable*.

(ii) If $\lim_{x \to c+} f(x) \neq \lim_{x \to c-} f(x)$, then f has a *jump* discontinuity at c.

(iii) If $\lim_{x \to c} f(x)$ does not exist for some other reason, then the discontinuity at c is called an *essential* discontinuity.

We are now equipped to characterize the set D_f for an arbitrary monotone function f.

Exercise 4 Let $f : \mathbf{R} \to \mathbf{R}$ be increasing. Prove that $\lim_{x \to c+} f(x)$ and $\lim_{x \to c-} f(x)$ must each exist at every point $c \in R$. Argue that the only type of discontinuity a monotone function can have is a jump discontinuity.

Exercise 5 Construct a bijection between the set of jump discontinuities of a monotone function f and a subset of $\mathbf{Q}$. Conclude that D_f for a monotone function f must either be finite or countable, but not uncountable.

D_f for an Arbitrary Function

Recall that the intersection of an infinite collection of closed sets is closed, but for unions we must restrict ourselves to *finite* collections of closed sets in order to ensure the union is closed. For open sets the situation is reversed. The arbitrary union of open sets is open, but only finite intersections of open sets are necessarily open.

Definition 3 A set that can be written as the countable union of closed sets is in the class F_σ.

To this point, we have constructed functions where the set of discontinuity has been $\mathbf{R}$ (Dirichlet's function), $\mathbf{R}\setminus\{0\}$ (modified Dirichlet function), $\mathbf{Q}$ (Thomae's function), $\mathbf{Z}$, and $(0, 1]$ (Exercise 1).

Exercise 6 Show that in each case we get an F_σ set as the set where each function is discontinuous.

The upcoming argument depends on a concept called α-continuity.

Definition 4 Let f be defined on $\mathbf{R}$, and let $\alpha > 0$. The function f is α-*continuous at* $x \in \mathbf{R}$ if there exists a $\delta > 0$ such that for all $y, z \in (x - \delta, x + \delta)$ it follows that $|f(y) - f(z)| < \alpha$.

The most important thing to note about this definition is that there is no "for all" in front of the $\alpha > 0$. As we will investigate, adding this quantifier would make this definition equivalent to our definition of continuity. In a sense, α-continuity is a measure of the variation of the function in the neighborhood of a particular point. A function is α-continuous at a point c if there is some interval centered at c in which the variation of the function never exceeds the value $\alpha > 0$.

Given a function f on $\mathbf{R}$, define $D_{f,\alpha}$ to be the set of points where the function f fails to be α-continuous. In other words,

$$D_{f,\alpha} = \{x \in \mathbf{R} : f \text{ is not } \alpha\text{-continuous at } x\}.$$

Exercise 7 Prove that, for a fixed $\alpha > 0$, the set $D_{f,\alpha}$ is closed.

The stage is set. It is time to characterize the set of discontinuity for an arbitrary function f on $\mathbf{R}$.

Theorem 2 *Let $f : \mathbf{R} \to \mathbf{R}$ be an arbitary function. Then, D_f is an F_σ set.*

Proof. Recall that
$$D_f = \{x \in \mathbf{R} : f \text{ is not continuous at } x\}.$$

Exercise 8 If $\alpha_1 < \alpha_2$, show that $D_{f,\alpha_2} \subseteq D_{f,\alpha_1}$.

Exercise 9 Let $\alpha > 0$ be given. Show that if f is continuous at x, then it is α-continuous at x as well. Explain how it follows that $D_{f,\alpha} \subseteq D_f$.

Exercise 10 Show that if f is not continuous at x, then f is not α-continuous for some $\alpha > 0$. Now explain why this guarantees that
$$D_f = \bigcup_{n=1}^{\infty} D_{f,\frac{1}{n}}.$$

Because each $D_{f,\frac{1}{n}}$ is closed, the proof is complete. ∎

To fully appreciate this result, it would be useful to have an example of a subset of $\mathbf{R}$ that is not an F_σ set.

Exercise 11 Figure out how to use the Nested Interval Property to prove that if
$$\{G_1, G_2, G_3, \ldots\}$$
is a countable collection of dense, open sets, then the intersection $\bigcap_{n=1}^{\infty} G_n$ is not empty.

(b) Use (a) to argue that it is impossible to write $\mathbf{R} = \bigcup_{n=1}^{\infty} F_n$, where for each $n \in \mathbf{N}$, F_n is a closed set containing no nonempty open intervals.

(c) Now show that the set $\mathbf{I}$ of irrationals cannot be an F_σ set.

A Continuous Nowhere-Differentiable Function

Exploring the relationship between continuity and differentiability has led to both fruitful results and pathological counterexamples. The bulk of our discussions to this point have focused on the continuity of derivatives, but historically a significant amount of debate revolved around the question of whether continuous functions were necessarily differentiable. Now we know that continuity is a requirement for differentiability, but, as the absolute value function demonstrates, the converse of this proposition is not true. A function can be continuous but not differentiable at some point. But just how nondifferentiable can a continuous function be? Given a *finite* set of points, it is not difficult to imagine how to construct a graph with corners at each of these points, so that the corresponding function fails to be differentiable on this finite set. The trick gets more difficult, however, when the set becomes infinite. For instance, is it possible to construct a function that is continuous on all of $\mathbf{R}$ but fails to be differentiable at every rational point? Not only is this possible, but the situation is even more delightful. In 1872, Karl Weierstrass presented an example of a continuous function that was not differentiable at *any* point. (It seems to be the case that Bernhard Bolzano had his own example of such a beast as early as 1830, but it was not published until much later.) Weierstrass actually discovered a class of nowhere-differentiable functions of the form

$$f(x) = \sum_{n=0}^{\infty} a^n \cos(b^n x)$$

where the values of a and b are carefully chosen. Such functions are specific examples of Fourier series. The details of Weierstrass' argument are simplified significantly if we replace the cosine function with a piecewise linear function that has oscillations qualitatively like $\cos(x)$.

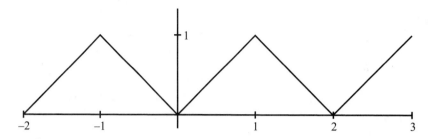

Figure 2.2.1. The function $h(x)$.

Define
$$h(x) = |x|$$
on the interval $[-1, 1]$ and extend the definition of h to all of **R** by requiring that $h(x + 2) = h(x)$. The result is a periodic "sawtooth" function (Fig. 2.2.1).

Exercise 12 Sketch a graph of $(1/2)h(2x)$ on $[-2, 3]$. Give a qualitative description of the functions
$$h_n(x) = \frac{1}{2^n} h(2^n x)$$
as n gets larger.

Now, define
$$g(x) = \sum_{n=0}^{\infty} h_n(x) = \sum_{n=0}^{\infty} \frac{1}{2^n} h(2^n x).$$

The claim is that $g(x)$ is continuous on all of **R** but fails to be differentiable at any point.

Continuity

The definition of $g(x)$ is a significant departure from the way we usually define functions. For each $x \in \mathbf{R}$, $g(x)$ is defined to be the value of an infinite series.

Exercise 13 Fix $x \in \mathbf{R}$. Argue that the series
$$\sum_{n=0}^{\infty} \frac{1}{2^n} h(2^n x)$$
converges absolutely and thus $g(x)$ is properly defined.

Exercise 14 Taking the continuity of $h(x)$ as given, reference the proper theorems that imply that the *finite* sum
$$g_m(x) = \sum_{n=0}^{m} \frac{1}{2^n} h(2^n x)$$
is continuous on **R**.

This brings us to an archetypical question in analysis: When do conclusions that are valid in finite settings extend to infinite ones? A finite sum of continuous functions is certainly continuous, but does this necessarily hold for an infinite sum of continuous functions?

2.2 Making the Epsilons Matter

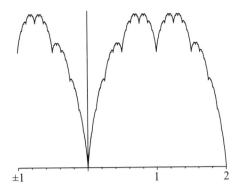

Figure 2.2.2. A sketch of $g(x) = \sum_{n=0}^{\infty}(1/2^n)h(2^n x)$.

Exercise 15 What theorem can we use in this case to prove that the infinite sum

$$g(x) = \sum_{n=0}^{\infty} \frac{1}{2^n} h(2^n x)$$

defines a continuous function on **R**.

Exercise 16 (a) How do we know that the function $g(x)$ attains a maximum value on the interval $[0, 2]$? Determine this maximum value.

(b) Let A be the set of all points in $[0, 2]$ where g attains its maximum. Find one point in A.

(c) Is A finite, countable, or uncountable?

Nondifferentiability

The more difficult task is to show that g is not differentiable at any point in **R**. Let's first look at the point $x = 0$. Our function g does not appear to be differentiable here (Fig. 2.2.2), and a rigorous proof is not too hard. Consider the sequence $x_m = 1/2^m$, where $m = 0, 1, 2, \ldots$.

Exercise 17 Show that
$$\frac{g(x_m) - g(0)}{x_m - 0} = m + 1,$$
and use this to prove that $g'(0)$ does not exist.

Any temptation to say something like $g'(0) = \infty$ should be resisted. Setting $x_m = -(1/2^m)$ in the previous argument produces difference quotients heading toward $-\infty$. The geometric manifestation of this is the "cusp" that appears at $x = 0$ in the graph of g.

Exercise 18 (a) Modify the previous argument to show that $g'(1)$ does not exist. Show that $g'(1/2)$ does not exist.

(b) Show that $g'(x)$ does not exist for any rational number of the form $x = p/2^k$ where $p \in \mathbf{Z}$ and $k \in \mathbf{N} \cup \{0\}$.

The points described in Exercise 18 (b) are called *dyadic* points. If $x = p/2^k$ is a dyadic rational number, then the function h_n has a corner at x as long as $n \geq k$. Thus, it should not be too surprising that g fails to be differentiable at points of this form. The argument is more intricate at points between the dyadic points.

Assume x is *not* a dyadic number. For a fixed value of $m \in \mathbf{N} \cup \{0\}$, x falls between two adjacent dyadic points,
$$\frac{p}{2^m} < x < \frac{p+1}{2^m}.$$
Set $x_m = p/2^m$ and $y_m = (p+1)/2^m$. Repeating this for each m yields two sequences (x_m) and (y_m) satisfying
$$\lim x_m = \lim y_m = x \quad \text{and} \quad x_m < x < y_m.$$

Exercise 19 (a) Without working too hard, explain why the partial sum $g_m = h_0 + h_1 + \cdots + h_m$ is differentiable at x. Now, prove that, for every value of m, we have
$$|g'_{m+1}(x) - g'_m(x)| = 1.$$

(b) Prove the two inequalities
$$\frac{g(y_m) - g(x)}{y_m - x} < g'_m(x) < \frac{g(x_m) - g(x)}{x_m - x}.$$

(c) Use parts (a) and (b) to show that $g'(x)$ does not exist.

Weierstrass' original 1872 paper contained a demonstration that the infinite sum
$$f(x) = \sum_{n=0}^{\infty} a^n \cos(b^n x)$$
defined a continuous nowhere-differentiable function provided $0 < a < 1$ and b was an odd integer satisfying $ab > 1 + 3\pi/2$. The condition on a is easy to understand. If $0 < a < 1$, then $\sum_{n=0}^{\infty} a^n$ is a convergent geometric series, and it is straightforward to conclude that f is continuous. The restriction on b is more mysterious. In 1916, G.H. Hardy extended Weierstrass' result to include any value of b for which $ab \geq 1$. Without looking at the details of either of these arguments, we nevertheless get a sense that the lack of a derivative is intricately tied to the relationship between the compression factor (the parameter a) and the rate at which the frequency of the oscillations increases (the parameter b).

Exercise 20 Review the argument for the nondifferentiability of $g(x)$ at nondyadic points. Does the argument still work if we replace $g(x)$ with the summation $\sum_{n=0}^{\infty}(1/2^n)h(3^n x)$? Does the argument work for the function $\sum_{n=0}^{\infty}(1/3^n)h(2^n x)$?

The Generalized Riemann Integral

If F is a differentiable function on $[a, b]$, then in a perfect world we might hope to prove that

(1) $$\int_a^b F' = F(b) - F(a).$$

Notice that although this is the conclusion of part (i) of the Fundamental Theorem of Calculus, there we needed the additional requirement that F' be Riemann-integrable. To drive this point home, we spent a considerable amount of effort constructing an example of a function that has a derivative that the Riemann integral cannot handle. The Lebesgue integral alluded to in our earlier conversations is a significant improvement. It can integrate the example we constructed, but ultimately it too suffers from the same setback. Not every derivative is integrable, no matter which integral is used.

What follows is a short introduction to the generalized Riemann integral, discovered independently around 1960 by Jaroslav Kurzweil and Ralph Henstock. This lesser-known modification of the Riemann integral can actually integrate a larger class of functions than Lebesgue's ubiquitous integral and yields a surprisingly simple proof of equation (1) above with no additional hypotheses!

2.2 Making the Epsilons Matter

Gauges and $\delta(x)$–fine Partitions

Let's quickly review the Riemann integral. Given a function $f : [a, b] \to \mathbf{R}$ and a tagged partition $(P, \{c_k\}_{k=1}^n)$, the *Riemann sum* generated by this partition is given by

$$R(f, P) = \sum_{k=1}^n f(c_k)(x_k - x_{k-1}).$$

Definition 5 Let $\delta > 0$. A partition P is δ-*fine* if every subinterval $[x_{k-1}, x_k]$ satisfies $x_k - x_{k-1} < \delta$. In other words, every subinterval has width less than δ.

This gives us the ingredients necessary to define what it means for a function to be Riemann-integrable.

Definition 6 A bounded function $f : [a, b] \to \mathbf{R}$ is *Riemann-integrable* with

$$\int_a^b f = A$$

if and only if, for every $\epsilon > 0$, there exists a $\delta > 0$ such that, for any tagged partition $(P, \{c_k\})$ that is δ-fine, it follows that

$$|R(f, P) - A| < \epsilon.$$

The key to the generalized Riemann integral is to allow the δ in the above definitions to be a *function of x*.

Definition 7 A function $\delta : [a, b] \to \mathbf{R}$ is called a *gauge* on $[a, b]$ if $\delta(x) > 0$ for all $x \in [a, b]$.

Definition 8 Given a particular gauge $\delta(x)$, a tagged partition $(P, \{c_k\}_{k=1}^n)$ is $\delta(x)$-*fine* if every subinterval $[x_{k-1}, x_k]$ satisfies $x_k - x_{k-1} < \delta(c_k)$. In other words, each subinterval $[x_{k-1}, x_k]$ has width less than $\delta(c_k)$.

It is important to see that if $\delta(x)$ is a constant function, then Definition 8 says precisely the same thing as Definition 5. In the case where $\delta(x)$ is not a constant, Definition 8 describes a way of measuring the fineness of partitions that is quite different.

Exercise 21 Consider the interval $[0, 1]$.
(a) If $\delta(x) = 1/9$, find a $\delta(x)$-fine tagged partition of $[0, 1]$. Does the choice of tags matter in this case?
(b) Let

$$\delta(x) = \begin{cases} 1/4 & \text{if } x = 0 \\ x/3 & \text{if } 0 < x \leq 1. \end{cases}$$

Construct a $\delta(x)$-fine tagged partition of $[0,1]$.

The tinkering required in Exercise 21 (b) may cast doubt on whether an arbitrary gauge always admits a $\delta(x)$-fine partition. However, it is not too difficult to show that this is indeed the case.

Theorem 3 *Given a gauge $\delta(x)$ on an interval $[a, b]$, there exists a tagged partition $(P, \{c_k\}_{k=1}^n)$ that is $\delta(x)$-fine.*

Proof. Let $I_0 = [a, b]$. It may be possible to find a tag such that the trivial partition $P = \{a = x_0 < x_1 = b\}$ works. Specifically, if $b - a < \delta(x)$ for some $x \in [a, b]$, then we can set c_1 equal to such an x and notice that $(P, \{c_1\})$ is $\delta(x)$-fine. If no such x exists, then bisect $[a, b]$ into two equal halves.

Exercise 22 Apply the previous algorithm to each half and then explain why this procedure must eventually terminate after some finite number of steps.

∎

Generalized Riemann Integrability

We now propose a new method for defining the value of the integral.

Definition 9 (Generalized Riemann Integrability) A function f on $[a, b]$ has *generalized Riemann integral* A if, for every $\epsilon > 0$, there exists a gauge $\delta(x)$ on $[a, b]$ such that for each tagged partition $(P, \{c_k\}_{k=1}^{n})$ that is $\delta(x)$-fine, it is true that

$$|R(f, P) - A| < \epsilon.$$

In this case, we write $A = \int_a^b f$.

Theorem 4 *If a function has a generalized Riemann integral, then the value of the integral is unique.*

Proof. Assume that a function f has generalized Riemann integral A_1 and that it also has generalized Riemann integral A_2. We must prove $A_1 = A_2$.

Exercise 23 Finish the argument by showing that $|A_1 - A_2| < \epsilon$ for an arbitrary $\epsilon > 0$.
∎

The implications of Definition 9 on the resulting class of integrable functions are far reaching. This is somewhat surprising given that the criteria for integrability in Definition 9 and Definition 6 differ in such a small way. One observation that should be immediately evident is the following.

Exercise 24 Explain why every function that is Riemann-integrable with $\int_a^b f = A$ must also have generalized Riemann integral A.

The converse statement is not true, and that is the important point. One example that we have of a non-Riemann-integrable function is Dirichlet's function

$$g(x) = \begin{cases} 1 & \text{if } x \in \mathbf{Q} \\ 0 & \text{if } x \notin \mathbf{Q} \end{cases}$$

which has discontinuities at every point of $\mathbf{R}$.

Theorem 5 *Dirichlet's function $g(x)$ is generalized Riemann-integrable on $[0, 1]$ with $\int_0^1 g = 0$.*

Proof. Let $\epsilon > 0$. By Definition 9, we must construct a gauge $\delta(x)$ on $[0, 1]$ such that whenever $(P, \{c_k\}_{k=1}^n)$ is a $\delta(x)$-fine tagged partition, it follows that

$$0 \leq \sum_{k=1}^{n} g(c_k)(x_k - x_{k-1}) < \epsilon.$$

The gauge represents a restriction on the size of $\Delta x_k = x_k - x_{k-1}$ in the sense that $\Delta x_k < \delta(c_k)$. The Riemann sum consists of products of the form $g(c_k)\Delta x_k$. Thus, for irrational tags, there is nothing to worry about because $g(c_k) = 0$ in this case. Our task is to make sure that any time a tag c_k is rational, it comes from a suitably thin subinterval.

Let $\{r_1, r_2, r_3, \ldots\}$ be an enumeration of the countable set of rational numbers contained in $[0, 1]$. For each r_k, set $\delta(r_k) = \epsilon/2^{k+1}$. For x irrational, set $\delta(x) = 1$.

Exercise 25 Show that if $(P, \{c_k\}_{k=1}^n)$ is a $\delta(x)$-fine tagged partition, then $R(f, P) < \epsilon$. Keep in mind that each rational number r_k can show up as a tag in at most two subintervals of P.
∎

2.2 Making the Epsilons Matter

Dirichlet's function fails to be Riemann-integrable because, given any (untagged) partition, it is possible to make $R(f, P) = 1$ or $R(f, P) = 0$ by choosing the tags to be either all rational or all irrational. For the generalized Riemann integral, choosing all rational tags results in a tagged partition that is not $\delta(x)$-fine (when $\delta(x)$ is small on rational points) and so does not have to be considered. In general, allowing for nonconstant gauges allows us to be more discriminating about which tagged partitions qualify as $\delta(x)$-fine. The result, as we have just seen, is that it may be easier to achieve the inequality

$$|R(f, P) - A| < \epsilon$$

for the often smaller and more carefully selected set of tagged partitions that remain.

The Fundamental Theorem of Calculus

We conclude this brief introduction to the generalized Riemann integral with a proof of the Fundamental Theorem of Calculus. As was alluded to earlier, the most notable distinction between the following theorem and the version we proved in class for the regular Riemann integral is that here we do not need to assume that the derivative function is integrable. Using the generalized Riemann integral, *every derivative is integrable*, and the integral can be evaluated using the antiderivative in the familiar way. It is also interesting to note that for the Riemann integral the Mean Value Theorem played the crucial role in the argument, but it is not needed here.

Theorem 6 *Assume $F : [a, b] \to \mathbf{R}$ is differentiable at each point in $[a, b]$ and set $f(x) = F'(x)$. Then, f has the generalized Riemann integral*

$$\int_a^b f = F(b) - F(a).$$

Proof. Let $P = \{a = x_0 < x_1 < x_2 < \cdots < x_n = b\}$ be a partition of $[a, b]$.

Exercise 26 Show that

$$F(b) - F(a) = \sum_{k=1}^{n} [F(x_k) - F(x_{k-1})].$$

If $\{c_k\}_{k=1}^n$ is a set of tags for P, then we can estimate the difference between the Riemann sum $R(f, P)$ and $F(b) - F(a)$ by

$$|F(b) - F(a) - R(f, P)| = \left| \sum_{k=1}^{n} [F(x_k) - F(x_{k-1}) - f(c_k)(x_x - x_{k-1})] \right|$$

$$\leq \sum_{k=1}^{n} |F(x_k) - F(x_{k-1}) - f(c_k)(x_x - x_{k-1})|.$$

Let $\epsilon > 0$. To prove the theorem, we must construct a gauge $\delta(c)$ such that

(2) $$|F(b) - F(a) - R(f, P)| < \epsilon$$

for all $(P, \{c_k\})$ that are $\delta(c)$-fine. (Using the variable c in the gauge function is more convenient than x in this case.)

Exercise 27 For each $c \in [a, b]$, explain why there exists a $\delta(c) > 0$ (a $\delta > 0$ depending on c) such that

$$\left| \frac{F(x) - F(c)}{x - c} - f(c) \right| < \epsilon$$

for all $0 < |x - c| < \delta(c)$.

This $\delta(c)$ is the desired gauge on $[a, b]$. Let $(P, \{c_k\}_{k=1}^n)$ be a $\delta(c)$-fine partition of $[a, b]$. It just remains to show that equation (2) is satisfied for this tagged partition.

Exercise 28 (a) For a particular $c_k \in [x_{k-1}, x_k]$ of P, show that

$$|F(x_k) - F(c_k) - f(c_k)(x_k - c_k)| < \epsilon(x_k - c_k)$$

and

$$|F(c_k) - F(x_{k-1}) - f(c_k)(c_k - x_{k-1})| < \epsilon(c_k - x_{k-1}).$$

(b) Now, argue that

$$|F(x_k) - F(x_{k-1}) - f(c_k)(x_k - x_{k-1})| < \epsilon(x_k - x_{k-1}),$$

and use this fact to complete the proof of the theorem.

∎

The impressive properties of the generalized Riemann integral do not end here. The central source for the material in this project is Robert Bartle's excellent article "Return to the Riemann Integral," which appeared in the *American Mathematical Monthly*, October, 1996. This article goes on to outline many other properties of this amazing integral. A more detailed development can be found in *Integral: An Easy Approach after Kurzweil and Henstock* by Rudolph Výborný and Lee Peng Yee or in a forthcoming book by Robert Bartle to be published by the American Mathematical Society.

2.2.3 The Task of the Educator

After many years of teaching introductory analysis, I collected my notes and wrote *Understanding Analysis*, an introductory text published in 2001 as part of Springer's UTM series. The preceding projects are edited versions of ones that appear in the book, and there are eight others in the text. Because the core content of the course is not heavily altered using this approach, one could surely use any number of excellent books and still incorporate the injunction to emphasize questions of analysis over questions of calculus.

Driven by the desire to teach well, all of us in this business continue to seek creative ways to motivate our students. In the spectrum of ideas that have been generated, I think it is fair to characterize the premise of this essay as rather modest but still important. To say it again, if we really want our students to be actively engaged in an analysis course, then we had better entice them with some firsthand exposure to the enigmatic delights that arise in the careful manipulation of the infinite. There are also some other very creative approaches to this course that should be considered. On the opening page of *A Radical Approach to Real Analysis*, David Bressoud begins with the following quote from Henri Poincaré: "The task of the educator is to make the child's spirit pass again where its forefathers have gone... In this regard, the history of science shall be our guide." Bressoud's remarkable book reads like a novel in some parts and effectively motivates new concepts by putting them into the historical context that generated the need for them in the first place. This idea is taken a step farther in *Analysis by Its History,* a fascinating text written by E. Hairer and G. Wanner that is even more ambitious in its attempt to recreate the theory of calculus and the 19th century transition to analysis "on period instruments."

Teaching a successful analysis course from any of these perspectives—and I have tried them all over the years—requires, as a first step, a deeply committed attitude on the part of the instructor. Nothing very good happens when we do not believe in the story we are telling or the way we have decided to tell it. In some cases the best advice is the most obvious. Let's remember why we were originally drawn to the subjects we teach and be sure to include in our courses the ideas that made us so passionate in the first place.

References

1. Stephen Abbott, *Understanding Analysis,* Undergradate Texts in Mathematics, Springer–Verlag, New York, 2001.
2. Robert G. Bartle, "Return to the Riemann Integral," *American Mathematical Monthly,* October, 1996.
3. David Bressoud, *A Radical Approach to Real Analysis,* The Mathematical Association of America, Washington D.C., 1994.
4. E. Hairer and G. Wanner, *Analysis by Its History,* Undergraduate Texts in Mathematics, Springer–Verlag, New York, 1996.
5. G.H. Hardy, *A Mathematician's Apology,* Cambridge University Press (Canto Edition), Cambridge, 1992.
6. Rudolph Výborný and Lee Peng Yee, *Integral: An Easy Approach after Kurzweil and Henstock,* Cambridge University Press, Cambridge, 2000.

Brief Biographical Sketch

Stephen Abbott is currently teaching in the mathematics department at Middlebury College. He has held visiting positions at St. Olaf College and the University of Virginia, where he earned his PhD in 1993. His published work includes articles in operator theory and functional analysis as well as collaborations in real analysis and probability. He also has a strong interest in the intersection of mathematics and the arts and recently has written and taught courses about science and theater.

2.3

Innovative Possibilities for Undergraduate Topology

Samuel Bruce Smith
Saint Joseph's University

2.3.1 Introduction

The development of topology ranks as one of the great success stories of twentieth century mathematics. While the precise definition of a topological space is not yet a full century old, the subject has become a core requirement for many branches of current mathematics research. From genetics to string theory to the social sciences, applications of topology are diverse and pervasive. In its own right, topology is a vital and ever growing area, comprising dozens of subfields and engaging hundreds of researchers around the world.

The status of the undergraduate semester-course in topology is, unfortunately, not quite so glorious. Introductory topology tends to be viewed as a course suitable primarily for students headed to graduate school. While there are many superb textbooks in the field, most pitch the subject at an advanced level, including far more material than is possible to cover in one semester. Ironically, the axiomatic rigor that makes topology a model and solid foundation for other fields is precisely the characteristic that makes it a difficult fit for the undergraduate curriculum.

In this paper, I hope to indicate how an introductory topology course can become an accessible and popular elective for math majors of various strengths and diverse goals. One of the great advantages of topology is the almost visual elegance of its formalism. By emphasizing this quality, a teacher can help students cope with the level of abstraction that is endemic to all theory courses. Of course, the subject matter of topology is its own greatest advertisement. By leading with the examples that have real geometric appeal, students can be motivated to tackle the more demanding aspects of the subjects. The structure of the course, moreover, should be sufficiently flexible to accommodate varied student needs. A topology course can function as a satisfying conclusion to course work in mathematics as well as a preliminary to graduate work. It can be the capstone of the pure mathematics curriculum and a starting point for independent research.

2.3.2 Motivating the Abstraction

Opening a topology text to a random page illustrates a basic point about the subject. You are likely looking either at a very intriguing picture or at a page of pure formalism: theorems, lemmas, proofs. This is a basic dichotomy of the field. In my experience, students are drawn to the subject matter of topology. The challenge is to help students do the hard work of mastering the formalism. An excellent way to succeed is to show students how elegant and satisfying the formalism can be especially when applied to a concrete and familiar problem.

Perhaps the most important example of the power and intuitive appeal of the formalism of topology takes the students back to first semester calculus. Every math major has nodded in agreement to the picture proofs of the Intermediate Value Theorem (IVT) and the Extreme Value Theorem (EVT). Interestingly, these theorems are rarely proven in calculus. When teaching calculus, I try carefully to prove each step in the chain of implications EVT $\Rightarrow$ Local Extrema Theorem $\Rightarrow$ Mean Value Theorem and make a serious effort to explain how the Mean Value Theorem leads to the Fundamental Theorem of Calculus. But I never try to prove the IVT or EVT. The reason, of course, is that these proofs properly belong to topology.

In a topology course, the proofs of the IVT and EVT reveal the basic orthodoxy of the subject. They indicate the power of the first definitions: open sets, continuity, and the crucial concept of topological invariance. The first step is to characterize the objects of the theorems, in this case intervals, in topological terms. This introduces the concepts of *connectedness* and *compactness*. The hard step is to prove that the defined properties actually do characterize the objects, that is, that the connected sets are precisely the intervals and the compact sets precisely the closed bounded intervals[1]. But now the power of the point of view takes over. By the definitions, these topological properties are preserved by continuous surjections; they are *invariants*. The proofs can, in fact, be visualized as commutative diagrams:

Theorem. (The Intermediate Value Theorem) If $f : \mathbb{R} \to \mathbb{R}$ is continuous and $I \subseteq \mathbb{R}$ is an interval then so is $f(I)$.
PROOF:

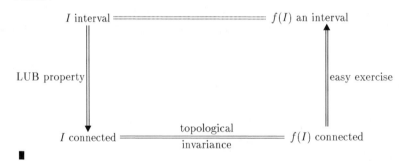

■

Theorem. (The Extreme Value Theorem) If $f : \mathbb{R} \to \mathbb{R}$ is continuous and $I \subset \mathbb{R}$ is a closed, bounded interval then so is $f(I)$.
PROOF:

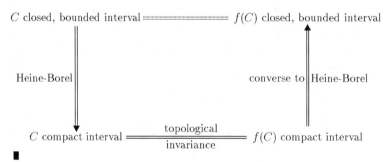

■

Figure 2.3.1. Visual Proofs of the IVT and EVT.

---

[1] As is often the case, the proof that the formalism really models the world is difficult!

2.3 Innovative Possibilities for Undergraduate Topology

The proofs of these important theorems make a compelling opening act to a course. For students intending to teach at the secondary level, the material is foundational. On the other hand, the notion of a topological invariant opens the door to advanced topics like the classification of spaces up to homeomorphism (the result $\mathcal{R} \not\approx [a,b]$ is proved while results like $S^n \not\approx \mathcal{R}^n$ make nice future goals). As with almost every topic in topology, there are many possible directions to pursue here and many intersections with other fields. Most importantly in terms of the course, these proofs can convince students of the elegance and the necessity of the formalism.

In a recent article in *The College Mathematics Journal* [1], Brenton and Edwards discuss how conceptual problems with sets become obstacles to understanding formal constructions like the quotient group in algebra. While students easily grasp the meaning of simple sets, e.g., sets of integers, they have problems with exotic sets like the set of cosets. Thus the quotient group in algebra remains mysterious due to the strange nature of its elements.

Topology is an excellent arena to work on tearing down this cognitive barrier. For example, consider, as the authors do, the case of $\mathcal{Z}_3$. Students are happy with the representation $\{0, 1, 2\}$ but understandably less so with the coset representation $\{0+, 1+, 2+\}$. Consider the possible representations of the unit circle S^1: algebraic, trigonometric, geometric, or as the identification space $[0, 1]/\{0 \sim 1\}$. Which is the most natural? The last option, being quite visual, is not so intimidating.

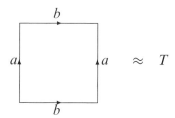

This picture introduces the quotient space, which is a good starting place for understanding more exotic sets. With some work, the topological isomorphism $S^1 \cong R/Z$ can be understood. The task is not any easier than it is in algebra but it might be, for some students, better motivated.

Next consider the torus T. The three-dimensional representation of T is manageable (and a good flashback to multivariable calculus). But the identification space representation below is perhaps even more natural.

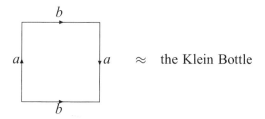

Routine exercises show that the topology is as expected. The product topology enters here via the homeomorphism $T \approx S^1 \times S^1$. This last fact, in turn, opens the door to topological groups. (Which spaces have this structure, how could we get an invariant, etc.?) Moreover, it is now a simple matter to obtain something strikingly different

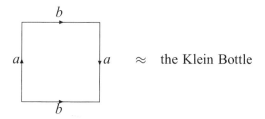

The abstraction now has become as much philosophical as mathematical. It is pretty clear what object we intend by the diagram. But what can we say about its existence? It can be represented (immersed) in three-dimensional space but it doesn't truly live (embed) there. To understand the Klein bottle mathematically we should use the formalism. The good news is that the formalism, in this case the quotient topology, is no more difficult than for the torus.

A topology course can feature the intuitive aspects of the subject without sacrificing the essential content. The preceding discussion hopefully serves to demonstrate how topics with concrete appeal can be used to introduce the key concepts of the course, providing a well-motivated path to the standard material. In the next section, I will offer some specific suggestions on how to organize such a course.

2.3.3 Structuring a Course

My basic goal in designing a topology course is to maximize the extent to which students discover the ideas of the course for themselves. Thus I emphasize problem sets over tests and in-class group work and student presentations over lectures. The advantage of this approach is clear: students feel ownership of ideas they have worked through themselves. The principal drawback is that less material can be covered than in a standard course, which might disadvantage the very best students. However, the structure of the course described below is flexible enough to allow the most capable students to work on suitably challenging problems without completely overwhelming the less advanced students. The key to this flexibility lies in the role of homework in the course.

Problem Sets

Working problems is, of course, essential for learning mathematics. I recommend making homework problems central to the course in terms of both weight and focus. In my experience, students feel less intimidated if homework comprises a substantial fraction of the total points since they have more control over their score on the problem sets than they do on a test. Of course, giving significant weight to homework also justifies assigning many problems.

I divide homework problems into two types: *practice problems*, which reinforce the basic concepts and are essentially routine, and *challenge problems*, whose resolution will be fundamentally more involved. Practice problems function as traditional homework problems; they have a due date and are graded and returned. Students are expected to attempt all of the practice problems.

Challenge problems, on the other hand, are elective: students can attempt those that interest them. Challenge problems have no due date but remain *open* until a correct solution has been presented to the class. Since it is not possible to have all the challenge problems presented in class, I also close challenge problems when they have been solved by a substantial number of students. Thus the challenge problems give the course the atmosphere of a research seminar. Moreover, the division of problems gives the students choices about what topics they study in-depth without sacrificing basic common knowledge. One of the pleasures of teaching topology is the great variety of possible topics and problems at all levels of difficulty. The following examples indicate some possible ways that problem sets can be used to engage students in the course and shed new light on other areas of mathematics.

Homework problems should be treated as a focal point of the class, not just as subsidiary exercises yielding results with little bearing on the theory developed in lecture. For example, consider the standard material on closure, interior, and boundary. A lecture on examples in Euclidean space motivates the definitions. Routine problems like $\overline{A} = A^\circ \cup \partial(A)$ can be divided up between lecture and the practice problems. Practice problems can also be formulated as "determine the closure and interior" of subsets of interesting topological spaces. Such problems give students early hands-on experience with exotic topologies. Challenge problems can be harder standard results such as $\overline{A \times B} = \overline{A} \times \overline{B}$ or (more challenging) $\overline{\prod A_i} = \prod \overline{A_i}$ or (very challenging) the Kuratowski 14-set problem [3, Exercise 20, p. 101].

Problem sets can be thematic. For example, a problem set on the Cantor set C is always popular with students. Practice problems can include the standard results (C is totally disconnected, perfect, etc.). Proving $C \approx \{0, 1\}^\omega$ is also not difficult and makes a good exercise in understanding the product topology. Challenge problems could include (the rather strange result) $C \approx C^\omega$ and (with hints) the various uniqueness results for C. This type of problem set can also open the door to independent research problems.

Homework topics can provide an instructive interaction with other areas of the undergraduate curriculum. An obvious example is topological groups. Here a couple of lectures can provide the background while practice problems can include proving properties like Hausdorff (assuming T_1) and homogeneity of coset spaces (a useful exercise with the quotient topology). Challenge problems can include regularity of G and of G/H or, for a concrete example, proving $GL(n, R)$ has exactly two path components.

The various incarnations of set theory in topology can be organized to provide a novel tour of Cantor's theory. For example, the space $\overline{S_\Omega}$ (in Munkres' notation), consisting of all countable ordinals union the first uncountable ordinal Ω in the order topology, is a rich source of counterexamples. Studying properties of this space in a problem set provides an opportunity to review and discuss the well-ordering theorem, the construction of the ordinals and the continuum hypothesis. Other possibilities in this direction are the theory of Baire spaces and the set theory involved in proving the Tychonoff theorem.

In-Class Assignments

Topology represents the context for one of the most famous pedagogical innovations of the twentieth century. The Moore method, as developed by R.L. Moore at the University of Texas, eliminates all texts and references from the course, forcing the students to truly discover the results of the subject for themselves. While the Moore method is probably not suitable for most undergraduate classes, it is possible to recreate part of the experience for students using in-class assignments. I set aside several classes each semester, timed to coincide with the start of a new topic, during which the students split into groups and work on problems. For example, after introducing the separation axioms (Hausdorff, regularity, normality etc.) the questions which axioms are *hereditary* (inherited by subspaces) and which are *productive* (passed to products) are natural and open ended. Resolving these questions is easy in the case of the Hausdorff axiom, hard but possible for regularity, and very difficult for normality. Thus these exercises give students a taste of two aspects of mathematical research: the consideration of propositions whose status is not known ahead of time and the great range of difficulty that similar-sounding statements can have. An alternate approach is to give each group a different, related problem (e.g., the many interrelationships among the countability and separation axioms) and have a volunteer from each group present its findings. In-class group work provides a lively alternative to lecture and an excellent opportunity for students to work on their topology language skills with help readily available.

Student Presentations

Every undergraduate mathematics major benefits from practice presenting mathematics to their peers. Unfortunately, student presentations are generally an inefficient use of class time since students tend to be a passive audience for their peers. Challenge problem presentations provide a partial solution to this problem. When a student volunteers to present a challenge problem the status of the problem is still open. Since the proposed solution has not yet been graded, the students themselves must judge the correctness of the presentation as well as understand the techniques used. I award points both to the presenter and to the audience. If the proof presented is incorrect then an audience member can score points for pointing out difficulties or, even better, for finding a fix to a problem. Since many students have attempted each challenge problem the students have real incentive to be an active audience. In this context, topology offers an advantage over other courses in that there are not only proofs to present but challenging yet accessible constructions as well. For instance, constructing an example of a path connected space which is nowhere locally connected makes for a nice challenge problem and presentation.

2.3.4 Independent Research Directions

An introductory course in topology is an excellent spring-board to undergraduate research. The breadth and pervasiveness of the field makes it easy to design independent study projects which exhibit a strong interplay of topology with other areas of the undergraduate curriculum. Such integrated projects have a dual benefit: they appeal to students whose ultimate interests lie outside topology proper and they offer a vista onto the world of mathematical research where there are no fixed boundaries between fields. Below are some possible directions for independent study and research in topology arranged roughly by their relationship to other (undergraduate-level) disciplines.

Analysis

After introductory courses in analysis and topology, there are a wealth of topics that can be pursued as independent studies. Examples include the theory of curves (Peano spaces and the Hahn-Mazurkiewicz Theorem), dimension theory, and fractals. Elementary functional analysis involves topology, analysis, linear algebra, as well as some elementary ring theory. (See [7] for an approach emphasizing the topological aspects of function spaces.) For students intrigued by the Cantor set, there are interesting advanced results related to its universal properties and its many generalizations.

Geometry

The classification of compact surfaces is a beautiful and accessible theorem with real geometric appeal. (See [3,4].) The notions of Euler characteristic, genus, and orientability are all illustrated by the theorem and can be pursued in more classical geometric contexts. Knot theory combines geometric appeal with connections to algebra and even physics. Geometric topics also occur in the theory of manifolds and in elementary differential topology.

Set Theory and Foundations

In addition to its many exotic set-theoretic examples (the ordinal space, the Tychonoff plank), advanced point-set topology offers a large supply of interesting problems in the intersection of topology and set theory. Consider, for example, the mathematics surrounding *Dowker's conjecture*. Since normality is not productive, it is natural to introduce stronger generalizations which are. The class of *binormal* spaces is characterized by the fact that the product with the unit interval $X \times I$ is normal for X binormal. The usual problem of finding a distinguishing example leads, in turn, to *Dowker's conjecture*: that, in fact, $X \times I$ is normal for every normal space X. M.E. Rudin proved that Dowker's conjecture is unprovable using the axioms of set theory but the question of independence remains open. (See [5] for a relatively recent reference on current research.) An independent research project on Dowker's conjecture thus involves both interesting advanced topology and modern set theory.

Combinatorics

For students interested in combinatorics, there are many possible projects with topological flavor. Compactifications are used in Ramsey theory, surfaces in graph theory, and simplicial complexes in the theory of order. An introduction to the theory of arrangements makes a great research project with a nice interplay of topology and combinatorics.

Algebra

The first elements of algebraic topology (homotopy, the fundamental group, covering spaces) make a natural sequel to an introductory topology course. Advanced topics include the Galois correspondence between the subgroup lattice of $\pi_1(X)$ and covering spaces of X, free groups via the fundamental group, and the Seifert-Van Kampen Theorem. Other topics in the intersection of algebra and topology include topological groups, transformation groups, and topological rings. Determining the structure of the various equivalence groups (homeomorphism, homotopy self-equivalence, etc.) of a topological space represents a fundamental and difficult problem in topology. However, aspects of this problem can be pursued as undergraduate research.

Computer Science

Effective computation in topology and its relationship to functional programming is an area of active current research. For example, in [5] the authors provide an advanced but readable account of a program named *Kenzo,* which gives a solution to the computability problem for the first homotopy groups of a simplicial complex. A student knowledgeable about topology and programming could implement an algorithm such as this in some special cases.

2.3.5 Attracting Students

Thus far we have discussed the many possibilities for content in an undergraduate topology course and have skirted over the issue of how to populate such a course. Topology is an obvious choice for a student planning on doing graduate work in mathematics. But the preceding discussion suggests that a larger and more diverse group of students can and should be exposed to the subject. The question remains as to how to attract these students.

My results in this area have been decidedly mixed. My first attempts to offer a course in topology resulted in one or sometimes two very seriously interested students, with no hope of drawing the four or five more needed to actually run the course. This was a great development for initiating undergraduate research in topology but not for running a course.

My current approach is to try to draw students from a larger pool, particularly from the group of future high-school teachers. At a departmental seminar devoted to describing current electives, I give a brief talk to juniors and seniors, sketching the foundational role of topological ideas for calculus and emphasizing, as above, the power of the abstract perspective in this context. Many of these students, especially those with an interest in secondary education, are actively involved as tutors in calculus. They also have the mathematical maturity to appreciate the need for a rigorous proof even of extremely intuitive statements like the Intermediate Value and Extreme Value Theorems. I also describe some of the visual examples mentioned above and the connections between topology and the other areas of mathematics.

It is a simple fact of life that the appeal of topology as a subject must do combat with its deserved reputation for difficulty. As a result, many students decide that a topology course is not for them. However, this aggressive approach to marketing has yielded a small but sufficient audience for a course with representatives evenly mixed between future graduate students and future high school teachers. The idea for organizing a flexible and participatory course, as described above, then comes into play in satisfying the diverse needs and interests of these students.

2.3.6 Conclusion

As regards the undergraduate curriculum, topology suffers from an embarrassment of riches. There are far too many wonderful problems, examples, theorems, and applications for a single semester course. A topology course can be organized, however, to turn this breadth into an advantage, allowing students of varied interests and abilities to pursue different problems and paths. Moreover, the visual appeal and elegance of the subject can be used to help students master the considerable level of abstraction. Finally, topology offers a wealth of possibilities for capstone experiences and integrated undergraduate research.

References

1. L. Brenton and T. Edwards, "Sets of sets: A Cognitive Obstacle," *The College Mathematics Journal* 34 (2003) 31–38.
2. W. Massey, *Algebraic Topology: An Introduction*, Springer, New York, 1990.
3. J. Munkres, *Topology*, 2nd edition, Prentice Hall, New York, 1999.
4. J. Rubio, F. Sergerat, "Constructive Algebraic Topology," *Bulletin des Science Mathématiques,* 126 (2002) 389–412.
5. M.E. Rudin, "Dowker spaces," in *Handbook of Set-Theoretic Topology,* 761–780, North-Holland, Amsterdam, 1984.
6. J. Rutter, *Spaces of Homotopy Self-Equivalences: A Survey,* Lecture Notes in Mathematics 1662 Springer, New York, 1997.
7. G. Simmons, *Topology and Modern Analysis,* McGraw Hill, New York, 1963.

Brief Biographical Sketch

Sam Smith is an algebraic topologist and an associate professor of mathematics at Saint Joseph's University in Philadelphia. Sam received his PhD from the University of Minnesota in 1993.

2.4
A Project-Based Geometry Course

Jeff Connor and Barbara Grover
Ohio University

2.4.1 Introduction

In the Fall of 1997 Ohio University replaced its traditional "Foundations of Geometry" sequence with one in which the students develop their own sets of axioms and use them to establish some well-known results of plane geometry. By constructing their own axioms, the students gain a sense of both the source and the role of formal axiomatic systems. Since axioms are introduced and developed as needed, the students gain an appreciation of the significance of each axiom as it is added to the set of axioms.

As the students start the course by developing their own axioms, the course is not amenable to the traditional lecture approach of developing the material. The students develop their axiom systems while working in structured cooperative groups and making use of a variety of manipulatives and software programs during their discussions. By the end of the sequence, the students have addressed all of the concepts included in the traditional course and more. They also gain, in our belief, a deeper understanding of the material than would be developed in the traditional lecture style course.

2.4.2 The General Approach

The projects described in this paper were designed for a geometry course taken primarily by prospective middle or high school teachers. The major theme of the projects is to connect experience and abstract mathematics. The early projects are designed to give students experience in working with non-Euclidean geometries while exploring the validity of certain common sense propositions in these geometries. Later, once the students become familiar with models of these geometries, they are asked to develop an understanding of unfamiliar mathematics using those same models. A second theme is that the students learn mathematics using the tools they will eventually use in their professional lives. Since prospective teachers are very likely to use cooperative group work and technology in their own classrooms, they learn geometry using these pedagogical styles and learning aides.

Although the course described below was designed with prospective teachers in mind, the same approach can be used for other courses and other audiences. To find topics that will help the students move from familiar to unfamiliar material, it may be helpful to look to the history of the subject for project topics. As will become clear, this was the basis upon which some of the geometry topics were selected. Alternatively, one could look for a problem that captures the students' imaginations and have them develop the tools they need to solve the problem in a project. For instance, as a prelude to working with coupled differential equations, one could have the students work in cooperative groups to design a policy to control an animal population.

The above themes are consistent with the constructivist theory of learning. The projects described below are designed to either have students build abstract knowledge from their experience or, when necessary, generate experiences that will lead to a better understanding of the topic. The projects also conform to the "Necessity Principle" proposed by G. Harel: "Students are most likely to learn when they see a need for what we intend to teach them...", where the need noted here is an intellectual need [4]. In this case, the initial intellectual need we work to satisfy is the need of the prospective teachers to eventually present geometric arguments to their students. We also use software and manipulatives to create further intellectual needs by presenting geometries (e.g., spherical) where some of the standard facts of plane geometry are no longer valid.

The influence of Dr. David Henderson also needs to be acknowledged; our course revisions came after the first author attended a NSF-funded workshop "Experience and Geometry" organized by Henderson. One of the major themes of the workshop was that mathematics and human experience are intertwined. All of our projects reflect that theme. Although we do not develop geometry in the same manner Henderson does in his book [5], his philosophy had a direct impact on our work.

2.4.3 The Motivation for Change

Before discussing some particular projects used in the course, it is helpful to review some of the motivation for changing the structure of the course. The "Foundations of Geometry I, II" is a two-quarter sequence taken primarily by prospective middle and high school teachers. The catalog describes the course as follows:

> Introduction to axiomatic mathematics via two finite geometries and a variety of interpretive models. Develops plane Euclidean and non-Euclidean geometries in rigorous fashion from axiomatic approach.

Prior to Fall 97, the course was typically an axiomatic development of absolute geometry (which makes no assumption regarding the uniqueness of parallel lines) and Euclidean geometry, with a brief discussion of hyperbolic geometry using the Poincaré disc as a model.

The major failure of the traditional approach to the course and the deepest motivation for change, at least in the first author's eyes, was that it did not give the students the experience of doing mathematics. Using the traditional approach, most of the students anticipated and experienced the course as a tedious exercise in the memorization of proofs of obvious facts. Given that nearly all of the students in the course have only had a high school background in Euclidean geometry, they typically do not see the need to make a careful development of geometry without making any assumptions about parallels. The brief time spent with the Poincaré disc was insufficient to justify the effort spent on developing the results of absolute geometry, especially since the proofs of several theorems of absolute geometry can be considerably simplified by assuming the Euclidean parallel postulate. In addition, the students tended to rely on the instructor to develop the material and often did not appear to develop the intellectual skills needed to learn and understand any new mathematics they might be asked to work with after graduation. Two major goals of the revision were to give the students the experience of developing mathematics as a human endeavor and to give them the intellectual tools that they would need to learn any new and unfamiliar mathematics.

Another impetus for changing the course came from the National Council of Teachers of Mathematics (NCTM). The 1989 NCTM *Curriculum and Instruction Standards* [7] suggested that high school geometry should include spherical geometry and local axiomatic systems. Students also should experience experimenting with interactive computer software packages to observe relationships and then develop deductive arguments for their discoveries. In addition, high school students planning to attend college "should gain an appreciation of Euclidean geometry as one of **many** [emphasis added] axiomatic systems."

The 1989 standards also suggested that teachers use a variety of teaching strategies. In particular, "greater opportunities should be provided for small-group work, individual explorations, peer instruction, and whole-class discussions in which the teacher serves as moderator." The general tenor of these recommendations is continued through the NCTM's 1991 *Professional Standards* [8], the 1995 *Assessment Standards* [9] and the 2000 *Principles and Standards* [10].

The NCTM recommendations are consistent with the recommendations of other organizations. The National Educational Technology Standards (NETS) for Teachers state that teachers should be able to design "effective learning environments supported by technology" and "implement plans that include strategies for applying technology" [6]. The Conference Board of Mathematical Sciences issued a report on the Mathematical Education of Teachers in Fall, 2001 which recommends that "prospective teachers have mathematics courses that develop a deep understanding of the mathematics they will teach" and mathematics courses should "develop the habits of mind of a mathematical thinker and demonstrate flexible, interactive styles of teaching." [2]

2.4.4 Course Description

While both the traditional and revised geometry courses address similar material for the first part of the sequence, there are differences in how the material is developed and in the order in which the topics are presented. In the revised course, the material is developed over a sequence of projects. Each project consists of two or three progress reports, which require each group to write-up the results of their investigations, and a final report. The final report gives the students an opportunity to rework some of the results of the progress reports and to synthesize the results of the progress reports and lectures related to the project. Each student works as a member of the same cooperative group throughout a project.

At the time of the revision, we anticipated that the change in format would require substantially more class time to address the material usually covered in the traditional course. As a result we increased the number of contact hours per week from three to five. As it turns out, most of the material from the traditional course can be developed in the first three-quarters of the sequence.

Changes Related to Content

One of the main differences in the two versions of the course is that non-Euclidean geometry is introduced much earlier and the students work with models of these geometries much more extensively. This serves to motivate some of the more difficult results of absolute geometry. For instance, while it is fairly easy to establish that the angle sum of a triangle is 180° in Euclidean geometry, it is much more difficult to establish that the angle sum of a triangle is less than or equal to 180° in absolute geometry. Since this proof requires the use of the triangle inequality, this result occurs late in the traditional development of absolute geometry. As described below, the revised course introduces this result, but not the proof, early in the course along with the surprising (empirical) observation that in the Poincaré disc (and on the sphere) the area of a triangle is determined by its angle sum.

The other significant difference is that topics progress from the familiar, couched perhaps in unfamiliar settings, to the technical and unfamiliar. For instance, the first two projects are on area and the angle sum of a triangle, which were normally developed in the second half of the traditional course. In the revised course, many of the more technical topics that appeared early in the traditional course (such as the ruler postulate and the betweenness axioms) are developed later in the course, after the need for the topic has been established.

Another difference is that the revised course contains a significant amount of transformational geometry that was not included in the traditional course. This topic is developed using a manipulative (the MIRA) that allows students to perform reflections first of all with pencil and paper, then by using vectors and matrices, and finally from an axiomatic viewpoint. Recent versions of the course also include an introduction to taxi-cab geometry, a topic which appears in some secondary school geometry textbooks, and isometries (and hence congruence) in taxi-cab geometry.

The Use of Structured Cooperative Groups

One significant pedagogical change is that the content of the course is introduced via group projects. As a result, approximately 70% of the class time is spent having the class work in structured cooperative groups. The class is divided into structured cooperative groups for the duration of a project. Each group consists of three to four students and the instructor determines the membership of each group. The instructor usually tries to balance each group in terms of mathematical ability, writing skill, gender, and compatibility. This is, in part, to create an atmosphere in which students are encouraged to take intellectual risks with each other and in the class. Each member of the group is assigned one of the roles of facilitator, recorder, reporter, or checker. The facilitator keeps the group on task and controls the discussion, the recorder writes up the results of the group discussion, the reporter is responsible for creating the final report and reporting to the class and the checker makes sure that all of the group members are following the discussion. (A three-person group does not have a checker.) The reporter keeps her/his role throughout the entire project while the other roles are rotated for every progress report. The groups are rearranged at the end of each project; the general objectives are to allow each student to have the role of reporter at least once and to permit each student to work with a variety of other students in the class.

In order to encourage participation during group work, a short quiz is given following each progress report and the final report. The questions are based on the group's work; a correct answer is the result the group reported, even if the answer is mathematically incorrect. The total of the quiz scores is used as part of the group grade for the report. Since every member of the group has to have a good understanding of the content of the report in order to get a good grade on the quiz (and hence the report), each member of the group is encouraged to be actively involved in the project and to make sure that the other group members are involved. It also helps discourage a dominant group member from commandeering the project and turning in her/his work as the group's work.

The Use of Technology and Manipulatives

The other significant pedagogical change involves the increased use of dynamic geometry software and manipulatives to give the students experience working in Euclidean and non-Euclidean settings. Since the class meets in a computer lab, the students have constant access to these tools.

The students work with the Poincaré disc via the program NonEuclid. This program provides a software simulation of the Poincaré disc which allows the students to make a variety of geometric constructions and perform reflections over lines. It also allows them to measure angles, the length of line segments and the area of a triangle. Since the program does the required constructions and makes the measurements, the students can work in the Poincaré disc without knowing the Euclidean geometry underlying it. As a result, they can empirically verify that the Poincaré disc satisfies the axioms of absolute geometry and yet the angle sum of a triangle can be less than $180°$. (In fact, the betweenness axioms are introduced by having the students check that both the Geometer's Sketchpad and NonEuclid appear to satisfy these axioms.)

The students use Lenárt spheres to work in spherical geometry. These are clear plastic globes with a spherical straight edge and protractor. Students typically find it surprising that lines of latitude are not lines and that any two lines (great circles) must intersect. Students can draw and measure circles and angles, and directly observe the results of the curvature of the sphere. For instance, students can directly observe the failure of the exterior angle theorem and that the failure of the proof (given in a lecture) can be traced to the failure of the betweenness axioms in spherical geometry.

The students use Geometer's Sketchpad to work in Euclidean geometry. Even though they find little surprising about the geometry it represents, the students find the Sketchpad quite useful. In addition to its well-known construction tools, it also allows students to study transformations and isometries in taxi-cab geometry. Other tools include the MIRA, which allows the students to perform accurate reflections with pencil and paper.

Assessment

Each group submits written progress reports and a final report at the conclusion of the project. The instructor returns the reports along with a set of written comments. Typically, these comments address mathematical correctness and style, while discussing any interesting observations the students may have written into the report. Errors are classified as either minor or major. In the progress reports, points are not deducted for minor errors; this allows students to take some risks and the instructor can be critical without being punitive. In the final report, however, points are deducted for both minor and major errors. As mentioned above, part of the grade for both the progress reports and final projects is determined by the group's performance on a quiz. The written report typically counts for 60% of the score while the quiz accounts for the other 40%.

In addition to the group work, each student is assessed for performance on quizzes, homework and journal entries. There also are three traditional exams (two midterms and a comprehensive final). By traditional examinations we mean closed-book examinations that cover the material developed in class. Students typically are asked to state some definitions, to provide proofs of some of the results developed in the lectures and projects, and to prove something they have not seen before. These examinations are taken by individuals, not groups, during scheduled class time and are intended to help assess each individual student's understanding of the material. Group work accounts for 25% to 30% of the student's final grade; the remainder is based on individual work.

2.4.5 Some Sample Projects

The following projects from the first quarter of the course illustrate the above themes and course revisions. The first two projects are discussed in detail to give a sense of how the material interconnects and how student understanding of the material develops. The later projects are only briefly summarized, but they show how the course closes and reflects a gradual shift in emphasis in the projects. It is not necessary to develop the entire course using projects. It may be useful to develop just one or two topics using the project based approach. For instance, the angle sum project could be used after a lecture-based development of incidence geometry and the basic results on betweenness and plane separation.

First Project: Area

The first project is used to introduce the students to axiomatic systems by having them develop proofs for some facts that everybody knows: the formulae for the area of a rectangle, parallelogram and trapezoid. In this project, the students develop an axiom system that will allow them to justify the standard area formulas for these figures. The students also verify that The Geometer's Sketchpad (GSP) is a model of their axiom system and then show that only some of the axioms hold on the surface of a sphere. The project takes about four weeks and is divided into three progress reports and a final report as follows:

Progress Report 1 By first assuming the formula for the area of a square and then for a triangle, the students then describe a procedure for finding the area of a polygonal region. This usually takes the form of partitioning the region into either squares or triangles and then summing the areas of the figures in the partition. They then use one of these procedures to justify the standard formulae for the area of a triangle, square, rectangle and trapezoid.

The procedure for dividing the region into triangles is usually quite straightforward, but the attempt to partition a region into square subregions leads to technical difficulties almost immediately. For instance, how does one divide a $\sqrt{2} \times 1$ rectangle into squares? The students rarely introduce concepts from calculus and are sometimes surprised to find that they are reinventing the notion of the limit in some of their procedures. Another issue that comes up is whether the notion of equidistant should be included in the definition of parallel lines. This continues to be a point of discussion until the Poincaré disc, where parallel lines are not equidistant, is introduced in the second project.

The instructor comments on this report tend to focus on the more obvious gaps in the students' arguments. For instance, if a diagonal is used to divide a parallelogram into two triangles, then it is often assumed that the two triangles will have the same height. A more subtle point is justifying (or creating an axiom) supporting the claim that a polygonal region can be divided into triangular subregions.

Lecture on Axiom Systems In this lecture students are introduced to the idea of an axiomatic system and the concept of a model for a set of axioms. As part of this lecture, they are introduced to the axioms of incidence geometry and, as models of this axiom system, finite geometries. The material in this lecture is used as the basis for a set of homework problems.

Progress Report 2 Using their work from the first progress report, the groups develop a set of axioms for a theory of area along with definitions for rectangles and parallelograms. They also verify that GSP is a model of their axiomatic system and then prove some of the standard area formulae using their axiomatic system.

Verifying that GSP is a model of their axiom system leads to some good mathematical conversations. The usual form of this verification is to test a large collection of figures. For instance, a typical axiom is that the area of a triangle is "one half the base times the height". To verify this, the students construct a triangle and compare the result of "one-half the base times the height" to the area as given by GSP's area function. In order to take advantage of the dynamic feature of the software, the height should be constructed in such a way that one can move the vertices and have the height change along with the triangle and hence test the conjecture for a large collection of triangles. Quite often, students don't think of using the dynamic aspect of the software in this fashion. This leads to conversations about how one should construct figures so as to keep their defining properties while being able to move their vertices. For example, how can we construct a rectangle so as to be able to move any of three of its vertices and still have a rectangle? This can also lead to a discussion of the role that a definition plays in mathematics.

Students are also encouraged to identify the gaps in their arguments. This is where they can include ideas such as a closed figure, interior of a region, being between two points, length, algebraic operations with real numbers, and the like; concepts that are important but would take them away from developing a theory of area. Often these notions are described using a large set of undefined terms that come out of their discussions. As the course proceeds and a more sophisticated system of axioms is developed, many of these notions are made precise.

2.4 A Project-Based Geometry Course

The instructor comments on these reports tend to focus on the axioms and definitions given by the students. The students' definitions are often imprecise and notation is occasionally left undefined. Some concepts are also left out of their list of gaps or undefined terms. The comments also address the given proofs and any errors or missing steps in their proofs. Often, remarks address their experiments to verify that GSP satisfies their axiom system.

Following this progress report there is a class discussion where the class agrees on a set of axioms that will be used in the third progress report and the final report.

Progress Report 3 The students are introduced to spherical geometry and the groups use Lénárt spheres to do some experiments in spherical geometry. In particular, they explore the validity of their area axioms on the sphere and discover that they do not work as expected. In particular, the area of a triangle cannot be computed using the "one-half times base times height" formula because the result is dependent upon which side of the triangle one uses as a base. Also, that one cannot construct a square on the surface of a sphere comes as a surprise. (This often leads to revisiting the definition of a square.) They then modify their axioms and, using the modified set of axioms, they derive a formula for the area of a triangle on a sphere.

Final Report Using the class axiom system, the students write up their axiomatic systems for area in the plane, provide proofs of the standard area formulas, and prove a formula for the area of a triangle on a sphere. The students are, of course, supposed to use the instructor's comments on the progress reports in preparing the final reports. While points were deducted only for major errors on the progress reports, points are deducted for both minor and major errors on the final report.

Most of the major flaws in the arguments are corrected in writing the final report. There are still some difficulties in applying the axioms and some imprecision in the definitions. For instance, that a particular construction leads to a subdivision of a region is often justified via an axiom that asserts a subdivision exists and height is often defined to be a line segment instead of the length of a line segment (and not defined relative to a particular base). Most groups have made good progress in understanding what an axiom system is and in organizing their thoughts into proofs. Also, at about this point in the course, the student journal comments on creating proofs tend to turn from negative ("I can't do proofs.") to positive, indicating that they feel more comfortable with proofs and doing proofs [3].

Second Project: Angle Sums

This project also has the students prove a fact that everybody knows: the sum of the measures of the interior angles of a triangle is 180°.

Introductory Lecture This project starts with a lecture on the construction of line segments and angles via a brief introduction to the betweenness axioms and the axioms related to the measurement of angles. No proofs are given; the purpose of this lecture is to give the students a working vocabulary for the project.

Progress Report 1 The students are shown a physical demonstration that the angle sum of a triangle is 180° and then asked to use this as the basis for a proof. Typically, the students find they need to add an axiom to the effect that if a line is a transversal to two parallel lines, then it forms a pair of congruent alternate interior angles. This axiom in hand, the students can then show that the angle sum of a triangle is 180°. They also use GSP to find a formula for the angle sum of a polygon of a figure with n sides. They find that in GSP the $(n-2) \times 180°$ formula works for figures with a special shape (convex); as part of the report they need to define this shape and then provide a justification for their formula.

Lecture This lecture is a more rigorous version of the first lecture. The betweenness axioms are used to justify some properties of line segments, rays, and so forth. The purpose of the lecture is to review different methods of proof.

Progress Report 2 This progress report introduces the Poincare disc and the software program NonEuclid. The Poincaré disc is introduced as another model where the incidence axioms and the betweenness axioms are satisfied. The progress report is centered on using NonEuclid to explore statements equivalent to Euclid's Fifth postulate. Such statements might include determining whether, given a line and a point not on it, there is a unique line parallel to the given line passing through the given point. Another example might include deciding whether two parallel lines are equidistant from each other. They also are asked to observe that the exterior angle theorem, which is introduced as an axiom, is valid in both GSP and NonEuclid. The groups also verify experimentally that the angle sum of a triangle is less than 180° in the Poincaré disc. The groups usually do quite well on this project. There is often some difficulty with performing the required constructions, but these are usually handled in class before the report is collected.

Lecture on Euclidean and Non-Euclidean Geometries This lecture draws on much of the experience the students have had since the beginning of the quarter and is used to introduce the Euclidean and non-Euclidean geometries via parallel postulates. The history of geometry is reviewed, especially the development of hyperbolic geometry. Also, several statements equivalent to the Euclidean parallel postulate are discussed with the aid of GSP and the Poincaré disc.

Final Report The final report requires students to use the exterior angle theorem to construct parallel lines and the Euclidean parallel postulate to show that if a line is a transversal to two parallel lines, then the line forms a pair of congruent alternate interior angles. They are also asked to prove that the angle sum of a triangle is 180° in Euclidean geometry and the angle sum of an n-sided convex polygon is $(n-2) \times 180°$.

These reports are usually quite good. There is some difficulty with how to use the hypothesis that a parallel is unique in a proof, but most students eventually understand this. There is also some difficulty understanding the concept of convexity and using it in a proof. (This difficulty persists for quite a while.)

Further Projects

At this point, the students are about a third of the way through the two-quarter sequence. They are working with software models of Euclidean and hyperbolic geometry and a physical model of elliptical geometry. The students have become comfortable working in groups. Typically the groups no longer need an active facilitator or checker, even though we continue to assign the roles in the event one is needed. General comments and concerns about group work stop appearing in the student journals as it becomes the accepted way of working in the class. Except for clarification of mathematical issues, it sometimes feels as if the instructor is no longer needed in the classroom once the groups are engaged in a project.

The first quarter closes with a study of the criteria for congruence of triangles and an introduction to transformations in the Euclidean plane using the MIRA. The students develop rigorous proofs in absolute geometry that if side-angle-side is a criteria for congruence, then so are angle-side-angle and side-side-side. Using the Lenárt spheres and NonEuclid, the groups discover that AAA is a criteria for congruence on the sphere and in the Poincaré disc. Although they cannot prove these results, they are usually justified on the basis that the area of a triangle is determined by the sum of its angular measures.

Up until now, the students have only been proving results in absolute or Euclidean geometry. The work with the sphere and Poincaré disc has been primarily experimental and aimed at broadening their experience in geometry. As the course continues, the projects change their focus from developing definitions and theorems to applying the students' experiences in previous projects to unfamiliar mathematics. For instance, in an axiomatic approach to transformational geometry the students are given the definitions, axioms, and theorems that they are to prove with little or no motivational background. As part of the projects, they are asked to develop an interpretation of the abstract material in the context of GSP, the Poincaré disc and the sphere. Ultimately they show that any composition of reflections can be described as a composition of three or fewer reflections. They also study the types of motions that each type of geometry admits.

2.4.6 Student Reaction and Performance

The students appear to enjoy this method of developing the material and believe that they have had some experience in doing mathematics. The anonymous student evaluations given at the end of the course often contain statements to the effect that the course was interesting, that working on projects made the course "more understandable," that they were able to develop theorems on their own, that they had to do most of the thinking, and that "It helped me grasp the info better than if it was just lecture."

Student journal entries indicated that nearly all of the students enjoyed working in groups and using technology. An analysis of student journals showed a surprising lack of concern regarding group grades and fellow group members pulling their own weight. Most of the concern over these issues was expressed at the beginning of the course during the first year it was offered; since then it has rarely surfaced in journal entries. Student journals also indicate that working in groups is beneficial because it allows for brainstorming, peer instruction, group checks of proofs, and confidence building. The students also found the use of dynamic software and the Lenárt sphere to be helpful. The primary positive theme regarding the use of technology is that it provides a context to do explorations and build intuition with different geometries. In a set of interviews with students taking the class, the students interviewed had a difficult time determining whether the group work or the use of technology proved more helpful.

The revised course has been taught repeatedly by both authors. We believe the revised course creates a greater student understanding of the subject matter. The mathematical conversations we have with students are deeper and more productive. Students now use material from the geometry course in other courses. Finally, the students appear to follow lectures with greater comprehension than in a traditional course.

An analysis of six sets of final exams indicated that students tend to do as well on exams in the revised course as in the traditional course. Given a choice between answering a question on a topic developed in group work or a topic developed in lecture, the students tend to answer the question on the topic developed in group work, although performance is slightly better on topics developed in lecture. A more detailed presentation of these results appears elsewhere [3].

2.4.7 Some Things to be Aware of

It helps to keep a few things in mind when moving to a project-based format, especially when the projects are somewhat open-ended in terms of what is acceptable work.

To be effective, the instructor needs to have a solid understanding of the subject material and a reasonable knowledge of the students' backgrounds. He/she not only must know the subject well enough to realize what mathematical background is required for a project but also must be familiar enough with the students' mathematical backgrounds to know if the project is within their range. If the students do not have the required background, work on the project comes to a halt and they can get quite frustrated. For example, if the project requires an inductive argument and induction is not a part of the course, it is important that at least some members of each group have been exposed to induction arguments. Also, since the students will come up with a variety of approaches to a project, the instructor needs to know the subject well enough to be able to help a group work with an inefficient or inelegant approach so that it eventually leads to a solution. If an approach will not work, the instructor must be able to develop a line of questioning that will help the group understand why an approach should be abandoned.

It is important to be (emotionally) prepared for what the students cannot do, even if their mathematical backgrounds indicate that the work can be done. For instance, at the beginning of the geometry course, students still have difficulty parsing the hypothesis and conclusion of a conditional statement even though they have had a course that introduces them to proof. They often have surprising difficulty taking what appear to be small steps. For instance, even though they can apply induction to derive summation formulas, they have difficulty in using a similar approach to establish results such as the angle sum of an n-sided convex polygon is $(n-2) \times 180°$. In fact, most groups do not even recognize it as a candidate for an induction argument. Part of the difficulty is lack of experience on their part; the geometry course may be the first course that expects them to draw actively from material covered in their previous courses. The instructor also must be open to the possibility that the department's mathematics curriculum may not effectively promote creative mathematical thought. It also may not develop the connections between the content of different courses. Hence students may not achieve the expected level of understanding.

The instructor must be explicit about how the project is going to be graded and what constitutes acceptable work. It takes two or three progress reports before the students believe that there really are a variety of correct ways to approach a project. Unfortunately, what may appear to be the clearest of instructions to the instructor may suffer from ambiguity when read by the students. Try as they might, the students cannot read the instructor's mind; he/she should be ready to offer clarification of the instructions as needed. Detailed feedback to the groups regarding their work is essential to developing a shared understanding of expectations over time.

Given the above, most of the group work should be done in class. If the project is done outside of class, a group can spend an inordinate amount of time either pursuing an unproductive approach or being stuck on what the instructor thought was an obvious step. By being almost immediately available, a group's current conundrum can provide an opportunity to model a mathematical approach to the problem. And if the group is working in class, the instructor may be able to help them make a connection before they abandon a good approach to the problem.

One strategy that can ease the transition to group work is to ask the students to respond to a journal question such as: "To what extent are you satisfied that your group is working productively?" or "What concerns do you have about group work?" Student responses can alert the instructor to misconceptions about expectations or to unproductive behaviors of group members.

It takes a while for students to become accustomed to working in groups, so do not expect immediate success. In the geometry course, it takes about four to five weeks before the groups start working effectively. At the outset, the students do not understand that there is not a single correct approach to the material. Once they understand that the instructor is not looking for a particular solution, the groups get better at initiating and pursuing their own approaches to the problem. In addition, the first couple of weeks allow the instructor to assess the background of the students and to fine tune the projects to the class. (For example, add background lectures or adjust the problems in the progress reports.)

In the geometry course, five or six seems to be an optimal number of groups. The class needs to be large enough to generate a variety of approaches to a problem or project. For difficult problems, it helps the class as a whole when one group realizes how to do it. Just the knowledge that one group got it spurs the others on to their own solutions. If there are too few students in the class, say only enough for two or three groups, there doesn't appear to be the critical mass required to maintain momentum through the difficult parts of the project. If there are too many students, say eight or more groups, it is difficult to work with all of the groups during each class and the amount of grading becomes overwhelming.

The main difficulties involved in using group work are fairly generic in nature and will show up in almost any mathematics courses at the same level as the geometry course discussed here. However, there is one difficulty that may be peculiar to using technology in a geometry course. Students sometimes have difficulty creating objects satisfying particular properties that remain stable with respect to those properties when you move to a free vertex. There is a tendency to use the programs as drawing tools to make one figure as opposed to construction tools that generate a large set of figures for which a property is invariant. A classical example involves the construction of parallelograms or equilateral triangles. A typical drawing does preserve properties of the figure when a vertex is moved and thus it is difficult for the students to use the dynamic aspects of software programs such as GSP, Cinderella, or NonEuclid. It generally is difficult to get students to take full advantage of the dynamic aspects of any of the software packages used. Students sometimes will construct a triangle in order to check a proposition and then erase it in order to construct a different triangle to check the same proposition. It is only with some difficulty that they finally realize that they can drag one vertex to create a different triangle.

While the instructor plays a role in each group's work, it is important to wean the students from depending on the instructor for ideas and direction. It can be difficult for both the instructor and the students to have the instructor provide only the gentlest of hints (if any at all) on how to start a problem or, when asked to confirm whether a solution is correct (or not), to tell a frustrated group of students to come to its own conclusion. It is much easier for both parties if the instructor gives a direct hint ("Why don't you just try...") or verifies the correctness of the work. The instructor needs to exercise judgment when a group asks for help. While most of the time groups eventually will abandon unproductive approaches and come up with their own solution, sometimes a group really does need a direct hint to get on with the report. Or the instructor may have to intervene to prevent a group from moving in an unproductive direction. It helps if each group contains at least one risk taker who will help get the group working on the problem. Also, if a group continues to be dependent upon the instructor, it is worth reviewing the instructor-group interaction; the instructor could be inadvertently promoting the group's dependence upon the instructor.

It probably is clear from the description of the course that it is fairly demanding of the instructor. In addition to the in-class work, written comments on the progress and final reports are part of the dialog between the groups and the instructor. The reports, with comments, should be returned promptly so it is important to plan on when to collect them. It typically takes about 45 minutes to grade each progress report and another 30 minutes to write the quiz after all of the progress reports are done. This is in addition to grading any homework assigned and exams. There is some trade-off in that we only assign homework once every two to three weeks instead of weekly. In a class of 24 or so, the group projects take about as long to grade as a typical homework assignment in a traditional class.

By including a group component to the grade, the final course grades might not differentiate between students as well as grades based purely on individual work. The group work is generally good, hence the strong students still tend to get high grades. However, some of the weak students get higher grades than they might have ordinarily earned. For instance, an individual with good group work grades and F's on the exams and homework can get a D for the course. While it is a relatively rare occurrence, an occasional student may not be an active group member. However, such students also tend to do poorly on exams and homework and end up receiving the appropriate grade.

2.4.8 Conclusion

Even though the project based approach is more demanding of the instructor than the traditional lecture approach, we feel it is worth the effort. Students leave the course with more confidence in their understanding of the concepts and feel prepared to use them. More than a few students have entered the course hating geometry only to start looking forward to the day when they could teach it themselves. In the case of prospective teachers, the material has been developed in a fashion similar to the way they will be expected to teach it. They are more likely to use dynamic software and manipulatives in a way that will promote mathematical understanding in their students. Perhaps even more important, having had a taste of doing mathematics (and liking it), there is an increased chance that the prospective teachers will convey some of their enthusiasm to their students.

References

1. J. Castellanos, NonEuclid, *Interactive Constructions in Hyperbolic Geometry*, http://cs.unm.edu/~joel/NonEuclid/.
2. The Conference Board of the Mathematical Sciences, *The Mathematical Education of Teachers, Issues in Mathematics Education*, Volume 11, American Mathematical Society. Providence, RI, (2001).
3. J. Connor, "Using Technology and Cooperative Groups to Develop a 'Deep Understanding' of Secondary School Geometry" in *Proceedings of the Second International Conference on the Teaching of Mathematics*, John Wiley and Sons, Inc., 2002.
4. G. Harel, "Two dual assertions: The first on learning and the second on teaching (or vice versa)," *American Mathematical Monthly*, Vol. 105, No. 7 (1998), 497–507.
5. D. Henderson, *Experiencing Geometry on Plane and Sphere*, Prentice-Hall, Upper Saddle River, NJ, 1996.
6. International Society for Technology in Education, *Standards for Teachers*, Eugene, OR, 2000.
7. National Council of Teachers of Mathematics, *Curriculum and Evaluation Standards for school mathematics*, Reston, VA, 1989.
8. National Council of Teachers of Mathematics, *Professional standards for teaching school mathematics*, Reston, VA, 1991.
9. National Council of Teachers of Mathematics, *Assessment standards for school mathematics*, Reston, VA, 1995.
10. National Council of Teachers of Mathematics, *Principles and standards for school mathematics*, Reston, VA, 2000.

Brief Biographical Sketches

Jeff Connor is an Associate Professor at Ohio University. He received his PhD from Kent State University in 1985, working in the area of sequence spaces and functional analysis. He has recently developed an interest in how students learn to prove theorems, especially in geometry.

Barbara Grover is an Associate Professor at Ohio University. She earned her PhD at the University of Pittsburgh and held a research position for 6 years following the degree. While there she worked on two national grants, the QUASAR project and Thinking Mathematics. Her research interests relate to the professional development of inservice teachers and instructional approaches that are effective at the undergraduate level.

2.5

Discovering Abstract Algebra: A Constructivist Approach to Module Theory

Jill Dietz
St. Olaf College

2.5.1 Introduction

While module theory is not a regular part of a typical undergraduate mathematics curriculum, the topic provides an excellent opportunity to tie together fundamental courses in group theory and linear algebra. Many individual faculty members, whole departments, and consortia have addressed a common student perception that the 10 to12 courses forming a mathematics major are all fairly distinct from one another. The "seven into four" movement promoted by, among others, the United States Military Academy is a prime example of changing the curriculum in such a way that students see deep relationships among calculus, linear algebra, differential equations, multivariable functions, and so on. At the theoretical level, module theory allows students to use their background in axiomatic mathematics to see modules as generalizations of vector spaces, whose structure, under the right conditions, mimics the structure of finitely generated abelian groups.

Most undergraduate mathematics majors in the United States require essentially a one semester course in algebra, which typically covers the basics of groups, rings, and fields. Second courses in abstract algebra might go deeper into group theory (Sylow theorems, series, solvability), Galois theory, or cover special topics such as representation theory, simple groups, algebraic coding theory, etc. Rarely do students learn about modules until they are in graduate school.

I do not argue that teaching module theory is better than teaching Galois theory or representation theory to advanced undergraduates. I merely suggest that the topic deserves consideration at the undergraduate level. Moreover, I find module theory a perfect topic to treat from a constructivist, pedagogical philosophy.

Rather than engage in a philosophical discussion of the pedagogical theory of constructivism, this paper describes my experience using a kind of constructivist, or guided-discovery, approach in a senior-level mathematics theory course.

I teach a reasonably modern Abstract Algebra I course: using discovery exercises to generate conjectures, presenting theorems and proofs at the blackboard, encouraging discussion. Three of the five times I have taught a follow-up algebra course (which I will refer to as my *Modules Course*), I have employed an active learning method. It is not the Moore method or a typical constructivist approach, but as near a true discovery approach as I dare get.

In Section 2.5.2 of this paper is a discussion of the philosophy of, and ideas behind, active learning methods, including the Moore method, modified Moore method, and constructivism. The overall scheme of the Modules Course is outlined in Section 2.5.3 with a detailed example of a typical class day given in Section 2.5.4. A description of student interactions and requirements is in Section 2.5.5. Section 2.5.6 concludes the paper with a personal evaluation of the Modules Course.

2.5.2 Active Learning

By active learning I mean any pedagogical approach to education based on the philosophy that students learn best when they actively engage in constructing, discovering, or applying their own knowledge. An active classroom may not entirely reject the lecture format but at the very least it supplements and complements lectures with student-based activities.

In mathematics, especially at the college level, the active learning terms one most frequently hears are *Moore method*, *modified Moore method*, and *constructivism*. Below I will briefly describe my interpretations of these three terms. In addition, I will describe the guided-discovery approach I use in my Modules Course.

Moore Method

The Moore method (also known as the Texas method) was developed by Professor R.L. Moore, who spent most of his career at the University of Texas at Austin (1920–1968). The basic tenet of Moore's approach is that bright students can best learn mathematics by producing results themselves and by working as individuals in a highly competitive atmosphere. Moore self-selected his students, aiming for uniformity especially in their lack of knowledge of the topic at hand (often topology). Indeed, according to F. Burton Jones [3] (a Moore student as well as a Moore method proponent), Moore "aimed to have a class as homogeneously ignorant (topologically) as possible" and one in which "competition was one of the driving forces." Moore began the course by supplying his students with axioms, definitions, and a sequential list of theorems. The students proved the theorems themselves and constructed examples which both illustrated the theorems and showed the necessity of the hypotheses [3]. (For more information, see any number of papers, reflections and correspondences available at *The Legacy of R.L. Moore Project* web site [5].)

Moore required students to work by themselves with no access to outside help other than from Moore himself. Students presented their work at the blackboard, with Moore calling on the weakest students first. Generally, the class did not proceed until at least one of the students had successfully proved the particular theorem for which the class was responsible.

Converts to the Moore method use the technique (or modifications thereof) in nearly every course one might offer in a mathematics department. Physicists and other non-mathematicians have also found success using the Moore method in their disciplines.

Modified Moore Method

There seems to be no single version of the modified Moore method. Some professors allow the use of a computer to help develop intuition about a subject and generate conjectures [7]. Other faculty combine lectures with student presentations [4]. A third typical modification is to allow students to work in groups [1]. Moore believed that individual competition was key to classroom success, but others believe just as strongly that cooperation breeds a successful and enjoyable learning environment.

Moore's success is seen not only in the number of PhD mathematicians who began their mathematical careers in his courses, but in the number of converts to some variation of his method of teaching. Common criticisms of the Moore method are that the competitive atmosphere is not a good learning environment for many students and that the method filters out the brightest students at the expense of the rest.

Constructivism

Constructivism is an important, though controversial, theoretical perspective in modern education research that has its roots in philosophy and cognitive and developmental psychology. It is not my aim to engage in a discussion of the relative merits of trivial vs. radical vs. social constructivism (promoted, among others, by Piaget [6], von Glasersfeld [2], and Vygotsky [9] respectively). Nor will I present any mathematics education research into the merits or success of using constructivism in abstract algebra (see the RUMEC web site [8] (*Research in Undergraduate Mathematics Education Community*) for this type of research). Rather, I will briefly explain my interpretation of constructivism in general and the manner in which it is implemented in the Modules Course.

There is no single agreed upon method of applying constructivist ideals in the mathematics classroom. Common themes of those whose teaching is informed by a constructivist philosophy are (1) directing an active classroom that is not focused on lecturing at a blackboard, (2) using strategies that encourage reflection and help create or enhance students' understanding of mathematical concepts, (3) advocating a cooperative learning environment, and (4) respecting each student's individual learning techniques.

Implementation of constructivist ideas in mathematics classrooms comes in many forms such as the use of manipulatives in elementary education, graphing calculators in secondary education, and computers in higher education. It is not simply the use of a calculator or computer that makes a constructivist classroom, but the manner in which it is used. Students are encouraged to investigate open questions, generate conjectures, and suggest solutions and proofs.

Many teachers and faculty are devoted to constructivism, though its practice is not widespread at the college level. While utilizing activities in the classroom is enormously popular—there are hundreds of published examples of projects, labs, modules, activities, worksheets, etc., let alone unpublished examples—basing an entire course curriculum on the ideas of constructivism remains rare. Even rarer still is the implementation of non-Moore method constructivism beyond the first two years of a typical undergraduate mathematics curriculum. I am myself a devotee of constructivist philosophy but generally do not even attempt to run a course strictly based on constructivist principles. I do, however, dive in deep with my Modules Course.

Guided Discovery in the Modules Course

The Modules Course is based on a constructivist philosophy, though it differs from the usual constructivist approach in that the students truly discover and create their own mathematics. Like a Moore method course, the term begins by giving the students a set of axioms that define an algebraic object with which they are unfamiliar. The students analyze this object for the entire term. They ask questions, conjecture and prove theorems, and choose the paths the class will follow.

I use the term "guided discovery" to describe the pedagogical style employed in teaching the Modules Course. For the most part, the students construct their own knowledge and choose their own course of study, but I try to subtly guide them with questions, comments, and even facial expressions. If the class goes too far down a tangential path (e.g., What is there to say about all the homomorphisms from the group of integers mod n to itself?), I help them focus on a more narrow question (e.g., What kind of algebraic structure does $\text{Hom}_\mathbf{Z}(\mathbf{Z}_n, \mathbf{Z}_n)$ have?). If the class cannot see an important point because their choice of examples is too narrow, I suggest they investigate other examples. For instance, if the students believe that generating sets are the same as bases, then I ask them to consider $\mathbf{Z}_6$, as both a $\mathbf{Z}$- and $\mathbf{Z}_6$-module. I do not tell the students that $\mathbf{Z}_6$ cannot always be considered a free module, I merely suggest that they look at the example and let them discover the fact for themselves.

Where the Moore method has the professor motivate a certain topic and then ask the students to prove various explicitly stated theorems, I ask the students to write their own definitions, supply their own examples, and conjecture their own theorems. If 20 students conjecture 20 different theorems, then it is my job to ask them to pursue the most interesting conjectures. The most interesting conjectures, however, are certainly not equivalent to the provable conjectures. I allow my students to go down the wrong path, but I do not let them get too far afield.

Where more typical constructivist methods have the professor pre-design an activity based on the topic of the day, my modules classroom has a more impromptu feel. One moment the students are discussing module homomorphisms and the next they are talking about bases. Since the students determine to a large degree what they will study on any given day, it is nearly impossible to prepare activities for the class. Rather, I come loaded with dozens of examples that will illustrate many different aspects of module theory. If the students conjecture that all torsion-free modules are free, I ask them to consider the rationals as a module over the integers. If the students want to learn more about bases by actually computing some examples, I help them organize their investigative activities. I ask teams of students to study different families of examples and then bring the class together to discuss what each student and group has learned.

The students are very good at cranking out computations, assigning tasks within their small groups, and generalizing results. (For example, if two students have computed $\text{Hom}_\mathbf{Z}(\mathbf{Z}, \mathbf{Z}_3)$ and $\text{Hom}_\mathbf{Z}(\mathbf{Z}, \mathbf{Z}_4)$, the class will conjecture the structure of $\text{Hom}_\mathbf{Z}(\mathbf{Z}, \mathbf{Z}_n)$ in general.) But it is difficult for them to see more subtle things, such as the limitations of their selection of examples. (For instance, they might not notice that the only module examples they used for a particular activity are actually rings. Thus, they might mistakenly conjecture that $\text{Hom}_R(R, M)$ is always a ring.) A well-placed comment, question, or new example can provide clarification. (For example, I might ask "In the conjecture, can M be any R-module, or does it need to be a ring?" Alternatively, I simply ask the students to check their conjecture against the case where $R = \mathbf{R}^2$ and M is the set of points on the line $y = 2x$.)

Students work in groups of three or four of their own choosing for the duration of the term. For the most part, each group studies the same mathematical topics, though they might be studied from different points of view. For example, if the topic of the moment is homomorphisms of modules, one group might focus on all the different homomorphisms they can find between two particular R-modules. Another group tries to define a map between an R-module and an S-module. A third group decides to replicate the isomorphism theorems in the module setting. Different points of view lead to a variety of conjectures, so we spend some class time discussing each group's efforts.

Once the list of conjectures is whittled down to a manageable number, the students either prove a conjecture or find a counterexample. Unlike the Moore method, the class will proceed even if a conjecture is not resolved. Sometimes it takes only a day or two before a conjecture is resolved; sometimes it does not happen until the end of the semester. All conjectures that we agree to pursue are proved or disproved by the end of the term.

A typical class period ends with a list of conjectures, questions, and problems that each student or group must research outside of class. Often, an individual group is particularly interested in a peculiar problem, so it is responsible for considering an extra problem. A detailed example of a typical day is given in Section 2.5.4 below.

2.5.3 Course Scheme

After a bit of review, the course begins with the following question: "What happens if we replace the vector space axiom that requires an action of a *field* on an additive abelian group with the new requirement that there is instead a *ring* acting on an additive abelian group?" The question sets-up the class to (at least initially) study modules over rings as an analog to the study of vector spaces over fields. This, of course, is only one way to view a module, but it is a convenient way for the purpose of the pedagogical philosophy of the course.

There is no text for the course, so most students have no idea that they are studying modules; in fact, the term is never used. Moore explicitly told his students not to look at textbooks. I would do the same if the students ever showed an interest in searching the library.

The analogy with vector spaces takes the students pretty far. They can use their background to define modules (they come up with their own term for the new algebraic structure), submodules, homomorphisms of modules, quotient modules, etc. They find examples of everything and prove old theorems such as the isomorphism theorems (for modules), the kernel of a (module) homomorphism is trivial if and only if the homomorphism is one-to-one, and the like.

But the analogy with vector spaces gets students into a quandary too. They quickly realize that the notion of a basis does not carry over to the new module setting as easily as they had hoped. They bump into the notion of torsion almost by accident, and it takes them into a new realm of mathematics.

While the Modules Course begins by exploiting the analogy between modules and vector spaces, it ends by emphasizing the analogy between modules and abelian groups. In particular, I use the students' knowledge of the structure theorem for finitely generated abelian groups (which I will refer to as the FGAG theorem) as a road map to the more difficult structure theorem for finitely generated modules over principal ideal domains (which I will refer to as the M/PID theorem).

Students can easily define torsion (torsion elements, torsion modules, torsion submodules). They understand that as algebraists they are generally more interested in sets that have nice algebraic properties as opposed to those that do not. Thus, the students happily restrict themselves to looking at modules over integral domains so that the torsion submodule of a module really is a submodule.

On their own, students see that a structure theorem for finitely generated modules should mimic the FGAG theorem by splitting off free stuff from torsion stuff. It takes my interjection, however, to convince the students that working over PID's is the correct setting. They have a slight notion that one needs (finitely generated) torsion-free to imply free, but they have no idea that working over a PID will guarantee such a thing. This part of the course is a key place where I truly feel the need to guide my students rather than let them fumble around in the dark.

The course culminates in the M/PID theorem, but along the way we may get sidetracked by looking at all sorts of ideas unrelated to the theorem. One of my three classes spent quite a bit of time considering the structure of $\text{Hom}_R(M, N)$. Another class spent time considering multiple module structures on a given abelian group and whether or not one could define a homomorphism between M as an R-module and N as an S-module. These tangential issues are just as important as the M/PID theorem. I do not want to stifle the creativity of my class, but I push towards the theorem because it is an excellent place to end the course.

2.5.4 A Typical Day

In this section I give a detailed example of a single 85-minute class period that might occur during a term.

- Before class begins, the students have completed a homework assignment that grew out of the previous class discussion. In this case, they defined a cyclic module and constructed examples.
- (10 min) The class begins with one group's presentation of its definition of a cyclic module. If other groups have different definitions, they will be posted as well. Two standard student definitions for the R-module M to be cyclic are:

(1) $\exists m \in M$ such that $M = \langle m \rangle_1 = \{nm \mid n \in \mathbf{Z}\}$,

(2) $\exists m \in M$ such that $M = \langle m \rangle_2 = \{rm \mid r \in R\}$.

The first is based on students' knowledge of cyclic groups, the second takes the R-module structure into account.

- (5 min) All groups contribute examples based on their particular definitions.
- (10 min) To resolve the definition discrepancy, I ask the students if a cyclic module generated by an element should actually be a module. After quickly agreeing, I ask them to construct cyclic modules over $\mathbf{R}[x]$ using the two definitions. Eventually they will find that the cyclic module generated by $x \in \mathbf{R}[x]$ really is a module under the second definition, but not the first.
- (5 min) The $\mathbf{R}[x]$ example gets a student thinking about the differences among the various kinds of cyclic objects we have encountered in algebra. The student asks about the relationships among cyclic modules, cyclic groups and principal ideals. For example, are all principal ideals also cyclic submodules?
- (15 min) I suggest that the class break into their groups and ponder this question by examining cyclic sub-objects of $\mathbf{R}[x]$ considered as $\mathbf{Z}$-modules, as $\mathbf{R}$-modules, and as $\mathbf{R}[x]$-modules.
- (5 min) We take a break at about the 45 minute mark.
- (15 min) Groups report on the examples they computed and take notes on examples they had not considered. The question on the differences among various kinds of cyclic objects is not resolved at this point, but the issues raised by the question are more clear. Since I know this topic will not be resolved in a few minutes time, I find a way to move on to another topic.
- (5 min) I ask the class what else there is in the world besides cyclic objects. The students remember having studied groups that had more than one generator (most students have extensive experience with dihedral and symmetric groups), so the idea of generating sets comes to the table.
- (5 min) A student volunteers a definition of a finitely generated R-module, and a discussion ensues.
- (5 min) The generating set definition reminds one or more students about vector space bases. I tell the class that this is an excellent line of thinking.
- (5 min) Homework is assigned that must be completed before the next class. First, the students are to pursue the question about the differences among various cyclic algebraic objects. Second, they are to find examples of generating sets. Third, they are to define a basis of a module and find examples.

One can see from this description that I am not teaching the class so much as guiding it. I spend a lot of time suggesting a variety of examples that the students had not already considered as a way of illuminating thoughts and ideas. While I generally let the students' own questions determine the course of study, there are certainly times when I try to steer them towards an idea. For example, my question "What else is there in the world besides *cyclic* objects?" was carefully asked. The emphasis on cyclic keeps the students thinking about generators, so it is no surprise that the next idea that pops up is about generating sets. A different question (e.g., How many different isomorphism types of cyclic groups are there?) would have urged the students down a different path (e.g., When R is a principal ideal domain, a cyclic R-module is isomorphic to R/aR for some $a \in R$).

A more standard class on module theory would not address many of the questions on which my students sometimes spent significant amounts of time (e.g., defining a homomorphism from an R-module to an S-module). We may not cover as much material as other classes, but my students do manage to learn a significant amount and do so in a way that helps them understand the process of mathematics, not just the results.

2.5.5 The Students

Background

Students enter the Modules Course with a good undergraduate background in axiomatic mathematics. Minimally, they have taken Linear Algebra (including real vector spaces and linear transformations) and Abstract Algebra (including the basics of group theory, ring theory, and some field theory). They know that algebraic objects are defined via a set of axioms, and that tweaking certain axioms leads to different structures and different theories (e.g., abelian vs. non-abelian groups). They have dealt with sub-objects (subgroups, subrings), special sub-objects (normal subgroups, ideals), homomorphisms, quotients, etc. And they are familiar with a variety of examples of vector spaces, groups and rings.

Assessment

Requirements of the course include: classroom participation, homework, a journal, a final module paper, and a research paper that is written and orally presented. All work is done in groups of three or four. It is the students' responsibility to make sure that each group member is contributing to the workload fairly or deal with the consequences. Group grades are given, with slight modifications made at the end of the term.

Participation in class is clearly expected and is never an issue with the students; otherwise there would be no course. Often, a group has a spokesperson, but every member of a group is expected to contribute to the work.

Homework is really part of the participation component of the course. Students generate questions and conjectures throughout the term which they are to investigate both inside and outside of the classroom. Sometimes I ask them to do something explicit such as find five examples of modules with torsion; other times they are to come up with a question to ask about modules with torsion.

The journal is submitted every two weeks, and is intended to be an organized set of class notes. Given the nature of the course, lines of thought are not presented linearly nor even logically. Even my own notes are jumbled, so I ask the students to clean up their notes in order that definitions, examples, conjectures, theorems, and proofs are clearly marked. Responsibility for the physical task of writing the journal rotates through the group members.

The final paper on module theory is to be written as though it were a textbook chapter on module theory. Journal submissions become source material for the paper but a higher level of organization is added. The paper is the culminating product of the course, so much time and effort is put into its writing and rewriting.

For the research project, groups either truly research a question related to algebra (not necessarily module theoretic) or study a special topic such as Wedderburn's theorem on finite division rings or a tidbit from Galois theory. Results are written in a form expected by a mathematics journal, and presented orally during the last few days of the term.

Students' final grades are based mostly on their group's work on the two papers. Individual modifications to the group grade are based on both my and peer assessment of the particular individual's contribution to group work. Amazingly, measurements of individual contributions to the group effort submitted by each group member during the peer assessment process associated with each paper are virtually in agreement with one another. Thus, the measurements can be used to help tailor grades to the individual.

Student Interactions

Students work exclusively in their own groups during class time, though the groups do interact in order to challenge conjectures or clarify points made by others. In part they stick to their own groups because I often assign different tasks to different groups. On the other hand, there is a good deal of mingling among the whole class of students outside of the classroom when there is a common assignment to be completed. In every course I teach, I encourage students to work together both inside and outside of the classroom, but never is group work more successful than in the Modules Course.

Along with working together on mathematics, the students seek my help both in the classroom and in my office. Regardless of the location, I am always careful not to give too much away. Even in my office hours I continue to encourage students' self-discovery of the facts of module theory. Their questions are often met with suggested avenues of inquiry rather than with easy answers. For example, if we are the middle of class discussions about module homomorphisms I will not tell an inquiring student whether or not $Hom_R(M, N)$ is a group or an R-module, but I will suggest that the student investigate some examples then try to generalize or make conjectures based on the calculations. If the student comes back to me two days later with computations and conjectures in hand but is still confused, then I am more apt to provide an actual answer to the question.

Reactions

I had about 60 students spread among the three Modules Courses I taught using guided discovery. All 60 felt overwhelmed and lost at the beginning of the course ("Sometimes [we] were not always sure where we were" was one student comment). Many students are accustomed to doing mathematics, but very few are familiar with creating mathematics. Mathematics can be a frustrating endeavor for professional mathematicians, let alone undergraduates. A few weeks into the course, students figure out what I expect of them, how the course is being conducted, and the dynamics of their groups. Near the middle of the course they realize that they actually do know some mathematics and can use their foundation to generalize familiar ideas and create new mathematics. At this point they are comfortable asking questions and are unafraid of chasing what might end up being fruitless or simply uninteresting ideas. At the three-quarter mark we are deep into the M/PID theorem, which is challenging for the students, but they also understand that they are learning and creating substantial mathematics. The biggest challenge at this point is keeping the students focused. They begin to feel that they have learned a bunch of unrelated junk and they start to forget basic things such as the definition of a module. The journal requirement is intended to counteract this feeling, as is the final module paper. By the end of the term the students are exhausted, but in writing their papers they realize how much they have learned and created and how well ideas flow together and culminate in the M/PID theorem. The paper is an essential ingredient in the success of the course ("Being cut loose to work on the project ... was one of the most worthwhile things I've done as an undergraduate").

When the students come out of the course they have a better sense of what it is like to be a mathematician ("The format of the class ... provided insight into the methods of mathematical inquiry" and "I felt like a mathematician" were two student comments). I believe they are happy to have had such an experience ("It is amazing to find a math class that is so fun"), but they are equally happy that not all of their mathematics courses are conducted in this manner.

2.5.6 Conclusion

My own reaction to the course is that it is exhausting, challenging, and invigorating. It is much easier to have complete control over a class than to give up that control to 20 young undergraduates. It is challenging to make sure that the students learn some mathematics while trying to keep quiet about the paths of study they choose to take. It is invigorating to help students learn to create mathematics. Most mathematicians, including myself, did not get such an experience until beginning work on our doctoral dissertations. Most of our students will not get PhD's, so experiencing the guided discovery approach allows them one of their few opportunities to peek into the world of a mathematician.

I have thoroughly enjoyed my three experiences using guided discovery at the senior undergraduate level of mathematics, but I will not use the approach in every class I teach. The energy level required to conduct such a class successfully is prohibitive. I might feel otherwise if my class sizes were smaller than 20, but guiding 20 students (or five groups) on their own paths to discovery is like having five distinct classes to teach on top of an already demanding teaching load. Nonetheless, I will use guided discovery again in a senior seminar setting.

While the Modules Course and a Moore method topology course differ drastically in fundamental ways (small group work vs. individual work, choosing the course of study vs. proving pre-determined theorems), they share a crucial reliance on axiomatic subject matter. It is a student's background in axiomatic mathematics and her drive, in contrast to mathematical talent necessary for success in a Moore method course, that enables her to succeed in something like a Modules Course. More generally, I believe that discovery learning at the level of theoretical mathematics is only truly effective if students have a solid education in prerequisite courses.

References

1. J. Dancis and N. Davidson, "The Texas Method and the Small Group Discovery Method," retrieved from the Legacy of R.L. Moore Project web site at:
   http://www.discovery.utexas.edu/rlm/reference/dancis_davidson.html.
2. E. von Glasersfeld, *Radical Constructivism in Mathematics Education,* Kluwer Academic Publishers, Dordrecht, The Netherlands, 1991.
3. F.B. Jones, "The Moore Method," *American Mathematical Monthly,* 84 (1977) 273–278.
4. W.T. Mahavier, "A Gentle Discovery Method," *College Teaching,* 45 (1997) 132–135.
5. The Legacy of R.L. Moore Project web site, http://www.discovery.utexas.edu/rlm.
6. J. Piaget, *The Origins of Intelligence in Children,* International Universities Press, New York, 1952.
7. P. Renz, "The Moore Method: What Discovery Learning Is and How It Works," *FOCUS: Newsletter of the Mathematical Association of America,* August/September (1999).
8. Research in Undergraduate Mathematics Education Community web site, http://www.cs.gsu.edu/~rumec.
9. L.S. Vygotsky, "The Development of Scientific Concepts in Childhood," in R.W. Rieber & A.S. Carton (eds.), *The Collected Works of L.S. Vygotsky,* Plenum Press, New York, 1987.

Brief Biographical Sketch

Jill Dietz earned her PhD from Northwestern University. She is currently an Associate Professor in the Department of Mathematics at St. Olaf College. Her interests are algebraic topology, finite group theory, and promoting undergraduate research in pure mathematics.

Chapter 3

Papers on Special Topics

Introduction 113

3.1 The Importance of Projects in Applied Statistics Courses, *Tim O'Brien* 115

3.2 Mathematical Biology Taught to a Mixed Audience at the Sophomore Level, *Janet Andersen* 127

3.3 A Geometric Approach to Voting Theory for Mathematics Majors, *Tommy Ratliff* 133

3.4 Integrating Combinatorics, Geometry, and Probability through the
Shapley-Shubik Power Index, *Matthew J. Haines and Michael A. Jones* 143

3.5 An Innovative Approach to Post-Calculus Classical Applied Math, *Robert J. Lopez* 163

Introduction

The third chapter relates undergraduate mathematics to areas which were not an object of study just a few years ago. The paper by Timothy O'Brien of Loyola University Chicago discusses biostatistics courses that enroll both mathematics and biology majors. These courses use student projects to evaluate or limit the results of research papers in the biological sciences that use statistics as a tool. The paper by Janet Andersen of Hope College describes a team-taught sophomore level course in Biology and Mathematics. This course analyzes research papers that use matrices or differential equations in their development. In both cases there is a great deal of emphasis on student participation and presentations.

The subject of the next two papers in this chapter is voting theory, an ongoing area of mathematical research whose results are accessible to undergraduates. These two articles are a bit different from the rest of the papers in this volume in that the focus is more on the mathematical content of voting theory and a bit less on the approaches used to present this content to the students. These papers also serve in a certain sense as primers for both faculty and students in an area where there are no appropriate undergraduate texts available. In the first article, Tommy Ratliff of Wheaton College discusses the geometric framework underlying some of the recent results obtained in voting theory. The course, whose prerequisite is a course in discrete mathematics, makes active use of student readings, papers, and projects. One of its goals is to make students better aware of the choice procedures available to them when they make decisions. The other paper, by Michael Haines of Augsburg College and Michael Jones of Montclair State University, is directed toward instructors who want to incorporate results from voting theory into upper division mathematics courses. In addition to providing the basic background needed to access, for example, some of the material in the reference section, this paper also discusses how the material can be used in different courses.

In the fifth and last paper in this chapter, Robert Lopez, formerly of Rose-Hulman Institute of Technology and now with Maplesoft, discusses, in the context of a classical applied mathematics course, how a computer algebra system should be the working tool for teaching and learning in a variety of upper division courses. The article stresses the importance of integrating the algebra system into all parts of the course and not merely using it as an add-on to solve problems.

Taken together, these five papers offer an insight into some areas ripe for future development. They serve as a reminder that our goal is not only to improve the quality of our major classes but also to make our offerings attractive to people outside our major who need additional work in our area in order to make their own pursuits more productive. The end results will be beneficial not only to our students but also to our profession as a whole. We will attract more students, more interested students, in the process.

3.1

The Importance of Projects in Applied Statistics Courses

Tim O'Brien
Loyola University Chicago

3.1.1 Introduction

While in the past statistics courses may have emphasized formulae and summarizing data, the focus today is more on the importance of statistics in answering researchers' queries by obtaining essential information. As a result, today's students see statistics as an aid to the research process. For example, statistical methods associated with the field of bioinformatics have come into prominence over the past decade to provide biomedical researchers with the statistical tools necessary to detect patterns in very large genetic data sets similar to those resulting from the U.S. Human Genome Project. Modern statistics courses stress both the practical applications of statistical methods and the active participation of students in the learning process. Undergraduate and graduate programs in statistics, such as the ones recently revised at Loyola University Chicago, typically emphasize both the statistical applications in coursework and the involvement of students in the statistical consulting activities of faculty members. One tangible result has been the development of a sense of confidence on the part of students when tackling methodological challenges that go beyond the classical course in introductory statistics.

Like many U.S. universities, faculty at Loyola are involved in teaching an array of courses in theoretical and applied statistics. These include introductory courses for less technically oriented students and more mainstream undergraduate courses for statistics and biostatistics majors and minors. Many of these courses require student participation through projects, papers and/or class presentations. The focus of this article is to discuss some of the ways in which these activities help students understand the usefulness of statistical tools in a broad spectrum of fields. Specifically, the focus here is on the use of projects in Loyola's one-year biostatistics sequence (which is typically attended by premed students majoring in biology) and in follow-up courses in applied regression analysis, categorical data analysis, experimental design, statistical software packages, nonlinear modelling, and optimal experimental design. We provide several examples that illustrate how these projects have become an invaluable tool in the teaching of applied statistics and biostatistics courses and how they provide students with increased confidence by allowing them to obtain the tools necessary to master sophisticated statistical techniques.

3.1.2 An Introductory Biostatistics Course

The public understandably can become confused when studies such as the one described in Pope *et al.* [23] report an association between sulfur-oxide pollution and both lung cancer and cardiopulmonary mortality while other studies find no such association. Modern students of biostatistics are taught to critically examine the underlying statistical techniques used in these studies before considering the conclusions reached. Thus the fact that Pope *et al.* [23] adjusts for study biases by using an extension to the commonly-used Cox proportional hazards statistical survival model (Zar, [28]) to adjust for dependent observations lends significant credence to the paper's findings. Students now understand that merely applying a statistical technique without regard for necessary assumptions can easily lead researchers to dubious or incorrect conclusions.

In order to provide introductory biostatistics students with basic statistical tools, these courses typically cover an introduction to probability (including an appreciation of the relevance of the Central Limit Theorem), regression and correlation, one- and two-sample t-tests and generalizations to single- and multi-factor analysis of variance (ANOVA) and covariance, and an introduction to categorical data analysis that includes basic chi-square tests. Instructors in these courses usually find that they do not have enough time to cover very relevant intermediate-level topics such as odds-ratios and relative risk, logistic and non-linear regression, non-parametric methods, baseline-category logits, the proportional odds model, etc. As a result, introductory students often are left unaware not only of some of the limitations of and assumptions underlying introductory statistical methods but also of the relevant extensions of these methods provided in more intermediate-level courses. Equally disturbing is the (not infrequent) situation in which courses in statistical methods are taught by non-statisticians (such as biologists or psychologists), who may be unaware of these limitations and assumptions. For example, Samuels and Witmer [24] provide a 4×2 table of data relating pain relief (with levels none, some, substantial, and complete) to treatment received by the patient (drug A versus drug B) with the caveat that the usual chi-square test is inappropriate in this situation. Yet it is unclear how many non-statisticians would know the correct statistical technique to analyze these data. Statistical educators must do a better job of helping students develop skills beyond the level of the usual introductory course. We have found that class projects provide an excellent opportunity to do just that.

Loyola's introductory undergraduate biostatistics course is offered through the Department of Mathematics and Statistics and cross-listed in the Biology Department. The course has been redesigned with the idea of developing a student's critical eye through the use of course projects. This course is called STAT/BIOL 335 and usually enrolls about 40 to 50 students per semester. The course syllabus can be viewed on the Web at

```
http://www.math.luc.edu/~tobrien/courses/stat335/Fall-2003/syllabus.html.
```

Instructors of this course invariably point out the advantages and limitations of each of the techniques discussed in the course. As a result, students learn to be somewhat skeptical of studies employing dubious assumptions or involving only ten subjects in each treatment arm. Since time constraints do not permit exposure to many of the intermediate topics mentioned above, we have found that these subjects are well suited for students' class projects.

Class Projects

For their projects, students are given the choice of either (1) analyzing a sufficiently rich data set and writing up and presenting their findings to the class or (2) critiquing the statistical aspects of two research articles of their choosing from professional journals such as *The Lancet, New England Journal of Medicine,*

AIDS Journal, Ecology, and the like. The only requirement of the data set or research articles is that the statistical methods involved in the data set or article go beyond the level of the class, including topics from such fields as survival analysis, nonlinear or logistic regression, and repeated measures. In order to find a relevant data set or collection of articles, students first identify research fields of interest to them. This usually requires several meetings between the instructor and each student in order to get the students focused on a data set or articles at an appropriate level. Students then are required to obtain the tools needed to critique or implement the given statistical techniques by consulting intermediate biostatistics textbooks such as Zar [28] or the *Encyclopedia of Biostatistics* (Armitage and Colton [2]). Ultimately, students are required to write up their findings or critique in a three to five page paper.

We have found that in the early stages of a project instructors act as mentors only to the extent of helping students to identify relevant research articles or data sets or to find resources such as textbooks where the students can then learn about the statistical techniques employed in their articles or needed to analyze their data. After this initial phase, students learn via self-study the relevant statistical methodologies so that they can then critique the use of these methodologies in the chosen research articles. For example, a student may need to read through a textbook's chapters on bioassay analysis and mixed nonlinear regression. In student evaluations, students have commented how this latter phase has helped them develop confidence in their ability to further their own formal and informal study of intermediate and advanced statistical methods.

In their class papers, students first give a short summary of the goal of the research and the hypotheses studied in each article or provide the important characteristics of the data set they have analyzed. They then focus on the mechanics of the new statistical tools encountered and on the adequacy of their implementation in the article or data set. Students' projects and papers are then evaluated on their criticism of the statistical techniques employed in the two articles, their understanding of these techniques as evidenced by their criticism, and their writing style, including grammar and punctuation. Since undergraduate students rarely have access to rich data sets, our STAT 335 students almost invariably choose to critique research articles. Yet when the author offered a similar introductory biostatistics course to graduate students at Loyola's Medical School during the Spring, 1999 semester, the opposite situation occurred. These latter students opted to use sophisticated statistical techniques on the data sets provided by their research advisors. The class project helped show these students the usefulness of applying intermediate statistical techniques in their own research.

To illustrate some of the mechanics of the class project for this introductory biostatistics course, we now describe the projects of two undergraduate Biology students, Jennifer Huston and Nick Moisan.

Example: Jennifer Huston

For her class project and paper, Jennifer Huston critiqued the statistical techniques used in the article by Walrand *et al.* [27], which investigates the relationship between age and the ability to renourish the body, and by Lau *et al.* [17], which examines the relevance of mite and cat allergen exposure for the development of childhood asthma. The former article used a two-way ANOVA design and analysis for some response variables and a principal components analysis (a data-reduction technique) for other variables for a small rat study. Thus, as was the case for all students, it was necessary for Jennifer to learn the necessary assumptions required to use these advanced techniques. In her paper, she correctly pointed out the limitations (such as the inherent assumption of normality) and potential biases in their application in a study involving only 6 rats in one of the study groups.

For her second article, one of the response variables was ordinal in nature ("current wheeze," "wheeze ever," and "doctors' diagnosis of asthma"), so a multiple logistic proportional odds model regression was used. Once again Jennifer focused her comments on potential shortcomings of the use of this model in light of influential observations and potential outliers. Jennifer's project thus showed her the usefulness of statistical techniques in medical research and underscored the often-overlooked underlying model assumptions. Jennifer also mastered some rather sophisticated statistical techniques. In her course evaluation she mentioned that she now felt confident that by knowing where to obtain the necessary resources she could teach herself the statistical methodology useful in analyzing many medical studies.

Example: Nick Moisan

While the articles critiqued by Jennifer were relatively similar and typical of clinical and pre-clinical research, the articles examined by Nick Moisan provided a study in contrasts. Nick's first article, Krabbendam *et al* [16], which appeared in the *Journal of Neuropsychiatry*, examined whether a relationship exists between deficits in cognitive processing and the temporal and limbic volumes in the brains of humans. Typical of many articles appearing in psychology-related journals, this study based its conclusions on a small number of subjects and used a multivariate analysis of variance (MANOVA) design and several MANOVA analyses. In contrast, Nick's second article, Kernan *et al* [15], used a case-control study to test for a link between phenylpropanolamine (present in cough and cold remedies) and hemorrhagic stroke. This latter article appeared in the *New England Journal of Medicine*, and (as is often the case in medical research) employed logistic regression and odds ratios to draw inferences. Its conclusions were based on a very large study enrolling over 2100 subjects randomly selected throughout the United States. As a result of this project, Nick was able to see the wide application of statistical methods in diverse settings and to understand some of the subtle distinctions in the level of statistical sophistication in disciplines such as psychiatry and medicine. As was the case with Jennifer, Nick developed a great deal of confidence in being able to teach himself complex statistical techniques. Not surprisingly, both of these students continued their studies in applied statistics by taking additional statistics courses offered by our department. In addition, several Biology students with similar interests have decided to pursue Loyola's new minor in Biostatistics.

Both of these examples illustrate some of the benefits of using projects and papers in introductory biostatistics courses. Students invariably observe that statistical techniques are misused in otherwise prestigious research journals. Furthermore, these students also develop an important level of confidence in their ability to understand the necessary requirements and assumptions for statistical tests. They also learn effective communication skills through their written papers and/or their class presentations. Finally, they gain a sense of independence and confidence in their ability to locate resources, both on-line and in the university library, which further their understanding of advanced statistical techniques.

3.1.3 An Advanced Biostatistics Course

Responding to requests from introductory biostatistics students to offer a follow-up biostatistics course, Loyola's Department of Mathematics and Statistics offered an advanced-level biostatistics course during the Spring 2001 semester. This course focused on many of the theoretical and methodological aspects of the statistical techniques highlighted in the introductory projects and papers. These included the statistical techniques used in survival analysis, nonlinear and generalized linear regression, and clinical trials. The current class syllabus, notes and assignments can be viewed on the Web at

```
http://www.math.luc.edu/~tobrien/courses/ab/course-homepage.html
```

This course, also cross-listed with Loyola's Department of Biology and called STAT/BIOL 336, was attended by twelve students, six from the Biology Department and six from Math & Statistics, and presented the additional challenge of structuring a new course for a rather diverse group. The course was taught using class-notes based on material from Agresti [1], Bates and Watts [4], Davidian and Giltinan [7], Littell *et al* [18], McCullagh and Nelder [20], Pinheiro and Bates [22], Stokes *et al* [25], Venables and Ripley [26], Zar [28]. Material from Ewens and Grant [9], Harrell [13] and Johnson and Wichern [14] was added for the Spring 2003 version of the course. The Minitab®, SAS® and S-Plus® statistical computer packages were used in class handouts and by students in their assignments.

Class projects also were used in this course but, since the class-size was small, students were required to work in pairs in order to foster interpersonal communication skills. The student population was quite diverse in terms of mathematical sophistication, so these teams paired one quantitative student and one biology student. As in the introductory biostatistics course, class projects were used in this course to stretch students beyond the level of the course. With such a small group of students, each of the six student pairs was required to make a presentation to the class. Mentoring and assessment followed along the lines of the project for the introductory course, but part of the advanced class project grade also reflected the quality and clearness of the class presentation. The following example, which focuses on the statistical detection of the interaction of anti-HIV drugs, is typical of the projects from this course.

Example: Mike Evans and Bahram Patel

Since they were interested in HIV research, Mike Evans and Bahram Patel chose to examine the data provided in Machado and Robinson ([19], p 2304) to test the synergistic or antagonistic nature existing between the anti-HIV drugs AZT (Zidovudine) and ddI (Videx). For this study, the amount of the HIV-1 (strain LAV-1) virus present was measured by reverse transcriptase (RT) activity. With the help of the course instructor, these students used the SAS® statistical software package in a somewhat novel fashion to fit to these data the 5-parameter log-logistic dose-response (nonlinear) model

$$\eta = \frac{\theta_1}{1 + (\frac{z}{\theta_2})^{\theta_3}}$$

where $\eta = E(RT)$ is the expected amount of RT and where

$$z = AZT + \theta_4 ddI + \theta_5 \sqrt{\theta_4 * AZT * ddI}$$

is the effective dose of the anti-HIV drug. For this study, the key model parameter is θ_5, the so-called "coefficient of synergy." For this parameter, statistically significant negative values indicate antagonism and significant positive values indicate synergy of the study drugs. The instructor pointed out to these students that this model was applied in Gerig *et al.* [10] to the detection of the antagonistic joint action of similar compounds in the growth of cucumber seedlings and the model assumptions were subsequently verified for Gerig's cucumber data. When this model was fit to the HIV data, Mike and Bahram's preliminary results failed to detect significant synergy between these two drugs, a result which would then lead some researchers to conclude that the two drugs act independently. But upon inspection of the model residuals, the students noted that the variability in RT tended to decrease with η, and thus that one of the key assumptions (equal variances) was violated. When this heterogeneity of variance was then incorporated into the model by letting the variance be of the form $\theta_6 \eta^{\theta_7}$, the estimate of the coefficient of synergy became significantly positive. This led Nick and Bahram to conclude correctly that the two HIV drugs do indeed enhance one another.

Thus, even though the statistical methods employed in this project were rather sophisticated, Mike and Bahram observed first-hand the importance of checking the underlying model assumptions. They also learned something about the wide applicability of nonlinear models in biomathematical modelling. These students also correctly pointed out that the findings of the above study have led biologists to study the physiological mechanisms involved, which in turn would explain the synergistic effects of these two commonly used HIV drugs. Since modelling both the mean and the variance is becoming more and more necessary in bioassay studies, this example provided all the Advanced Biostatistics students with additional statistical tools.

It is worth noting that we have used class projects in the introductory biostatistics course to provide students with confidence in their ability to obtain the necessary knowledge and skills to criticize the application of statistical techniques performed by biomedical researchers. But we also have used class projects in more advanced courses to analyze medical data in a more sophisticated and novel manner so as to better test researchers' hypotheses.

3.1.4 Subsequent Applied Statistics Courses

In addition to the introductory and advanced biostatistics courses, we also have found that class projects are a useful tool in other intermediate and advanced applied statistics courses, highlighting the usefulness of projects for both non-major and major statistics courses. For example, we have found that courses in categorical data analysis (CDA), applied regression, experimental design, and statistical software packages can benefit from class projects, papers, and presentations as well. The following examples, which involve the application of statistical methods in fields as diverse as medicine, sports, and anthropology, illustrate this point.

Example: Dara Mendez

As both an employee of an area pharmaceutical company and a part-time Loyola student, Dara Mendez enrolled in a CDA course offered during the Fall 2000 semester both to further her knowledge in this area of great practical importance and to help her in her role as a pharmaceutical biostatistician. Once again, projects were used to push students to develop the skills required to master a topic beyond the level of the course. As a result, students' final class presentations were more pedagogical in nature. After being provided with the necessary resources, students required only minimal mentoring to accomplish their goals. Once again, they were required to write up their results in the form of a three to five page descriptive paper, which they then distributed to their classmates in conjunction with their 15-minute presentation on the given topic. Topics were chosen (jointly by instructor and student) from those related to the study of CDA but which were slightly above the level of the class. Some of the areas covered included bioassay, nonparametrics, sample-size determination, and pharmacokinetic mixed-effects modelling.

For her project, Dara chose to focus on the statistical methodology involved in quantile bioassay analysis. Using both the data provided in Stokes *et al.* ([25] pp 331-2) and some data from her work projects, Dana illustrated how statistical methods can be used to quantify the potency of an experimental drug relative to a standard one. As a former Biology student, Dara focused her class presentation on the implications of bioassay and the relative potency of drugs in pharmacology. This allowed her to provide both the required framework and larger picture to her classmates. As a result, Dara's classmates benefited from an informative presentation and learned just how prevalent these course methods are in the workplace. And Dara mastered the theoretical justifications and techniques for the methodology she had been applying on the job.

Example: Paul Bell

Ever interested in the use of statistical methods in sports, Paul Bell viewed the class project for his applied regression course during the Spring 2000 semester as an opportunity to use data that he had obtained to develop statistical models to predict the attendance at major-league baseball games in Chicago, Atlanta, and Oakland. For this course, students were required to obtain sufficiently rich data sets to analyze. They then wrote up their findings in a course paper and conveyed them to their classmates via a classroom presentation.

Paul's data was based on the 1999 baseball season and his multiple-regression models included variables such as the day of the week, weather outlook, game number, and the level of the opposing team. Paul then tested his models using 2000 attendance figures and all of his prediction intervals contained the actual reported attendance. The research for this course project earned Paul a Loyola Mulcahy scholarship/grant which enabled him to continue his research with the author. A description of these grants can be found on the Web at the website http://www.luc.edu/depts/prehealth/Mulcahy.htm. This grant covered the cost of obtaining his data as well as his travel expenses to Vancouver in June, 2001, to present his findings at the International Biometrics Society conference. As a masters level Statistics student, this experience has proven very beneficial to Paul's professional development.

This class project provided Paul with the opportunity to see how statistical methods can provide predictive models in a research area of special interest to him. His results have aided others interested in predicting attendance at major sporting events, such as stadium managers, area law-enforcement personnel, and mass-transit coordinators. Paul presented his class project findings to the larger Loyola community in a university seminar and has submitted his results for publication in the applied statistics journal *Chance*.

Example: William Burroughs

An undergraduate anthropology major, William Burroughs, enrolled in a course in statistical methods and software packages (STAT 303) during the Spring 2001 semester to learn how to use statistical methods to analyze anthropological data. He chose to examine a paleopathology data set provided by his advisor, Loyola Anthropology Professor Anne Grauer, for his class project and presentation. Projects were used in this class in much the same way as they were for the applied regression course described in the previous example. Students were required to obtain sufficiently rich data sets which they then analyzed and discussed in a class paper and presentation. This process again served to stretch students beyond the level of the material presented in the course.

Professor Grauer's data is related to human skeletal remains (skulls) from medieval England and was used to predict the incidence of a specific disease related to anemia as a function of the age of the subject at death. These data lend themselves to the use of the logistic function

$$E(p) = \frac{e^{\alpha+\beta x}}{1 + e^{\alpha+\beta x}},$$

which relates the expected probability of the disease at death (p) to the subject's age at death (x); here α and β are the model parameters to be estimated from the data. Of special interest to these researchers was the estimated age at which the expected probability of the disease is 50%, denoted x_{50}. After the model was fit to the data, William noticed a distinct pattern in the residual plot that indicated the inadequacy of this model. With some assistance from the instructor with regard to the statistical methods involved, William then wrote a computer program that used a Box-Tidwell transformation (Samuels and Witmer, [24] p 53) where $x = (age)^\varphi$ was used in the above model in place of x = age. This latter extension was validated since the subsequent residual plot showed the required random pattern. William's analysis highlighted the importance of this transformation since the estimate of x_{50} dropped from 31 years for the (incorrect) naive logistic model to approximately 20 years for the transformed logistic model.

Through his class project, William came to understand the importance of testing model assumptions and of extending existing statistical methods to fit real-life situations. In addition, while the original analysis performed by Professor Grauer, which involved a chi-square test, was very weak (in terms of statistical power) for this study, the application of the generalized logistic model was novel and provided a more direct answer to the important research queries. As the result of his class presentation, William's classmates understood the usefulness of sophisticated statistical techniques in the field of anthropology.

The class projects for this course involved somewhat more mentoring on the part of the instructor since the statistical prerequisites for the course were indeed modest (some exposure to basic statistics). In addition, the emphasis of the course was more on developing students' statistical programming skills and less on requiring that they master a subject area in applied statistics. All the same, the projects provided students with a sense of confidence and appreciation of the usefulness of applying statistical methodology to real-life data.

3.1.5 Independent Study Courses

As is often the case, class projects in introductory and intermediate courses spark student interest in furthering their studies in statistical fields involving their specific research interests. As a result, a number of follow-up independent study courses have been offered for the more advanced students. The following two examples discuss the development of independent study projects with advanced undergraduate and graduate students enrolled in Loyola's Department of Mathematics and Statistics. These examples illustrate how class projects at this advanced level typically entail assisting a faculty member in cutting-edge research in statistical theory and methodology.

Example: Lisa Leigh and Katie Hanrahan

The exposure to logistic and log-linear models in a basic course in categorical data analysis during the Fall 2000 semester sparked the interest of Lisa Leigh and Katie Hanrahan in the larger field of nonlinear regression methods. As a part of an independent study course, each student worked through the text of Bates and Watts [4] with the author during the first half of the Spring 2001 semester. During the second half of the semester, each student assisted in research by focusing on separate problems related to nonlinear regression. Lisa concentrated on using SAS® software to obtain curvature measures for Gaussian nonlinear models while Katie centered on extending these curvature measures to cover non-Gaussian nonlinear models such as the odds-ratio, relative-risk, and logistic models.

Each of these class projects (independent study courses) required a great deal of mentoring by the instructor (several hours per week over the course of the semester). Assessment and evaluation was based on the final reports prepared by each student, including the quality of the corresponding computer programs. The benefit to the instructor in terms of quality research was significant since Lisa's input helped with the results developed in our recent submission, Haines *et al.* [12], to *Statistica Sinica*. The work with Katie resulted in a presentation by the author at the XXXIIIemes Journees de Statistiques conference in Nantes, France in May, 2001.

Through these class projects, Lisa and Katie mastered the course material in applied nonlinear regression. They were also given the opportunity to gain hands-on experience doing cutting-edge research in statistical methods which has helped each of them identify fields of interest for future study and research.

Example: Paul Bell and Nick Pajewski

The use of existing data sets, such as the HIV data set discussed earlier, to fit linear and nonlinear statistical models naturally leads one to wonder whether a better-designed study could provide researchers with the same amount of information but with fewer experimental runs. This is precisely one of the major goals of the field of optimal experimental design, and follow-up projects (independent study courses) along these lines were proposed by Paul Bell and Nick Pajewski for the Spring 2002 semester.

With the author's guidance, Paul and Nick worked through the material presented in the optimal design textbook of Atkinson and Donev [3]. As with the previous example, this project required a great deal of mentoring on the part of the instructor. As examples were encountered in the textbook, Paul and Nick alternated in developing the necessary computer programs to obtain and verify the optimal designs and the design methodology using the SAS/IML® programming language. With the instructor, Nick and Paul then worked through the papers of Downing *et al.* [8] and Haines *et al.* [11] and wrote SAS/IML® computer programs to obtain the optimal designs for the models presented in these papers as well as for the logistic and the synergistic HIV models discussed above. Nick and Paul were evaluated on their mastery of the optimal design material, on the quality of their computer programs, and on their final reports.

This project served to unify the information learned in such diverse courses as applied linear and nonlinear regression, categorical data analysis, experimental design, and statistical software by providing practitioners with efficient design strategies. Since in some cases optimal designs with only half as many runs provided the same level of information as those actually used, Paul and Nick now understand the importance of a well-designed study in terms of cost-savings.

3.1.6 Evaluation and Assessment

The assessment of student projects is an ongoing process. In lower level courses, students are given milestones that they must meet if the project is to be accepted. For example, they must come up with five potential research articles to critique by a certain date. They must meet with the instructor by a certain (later) date to go through the papers and come up with the two articles to critique. And they must complete the three to five page paper by a certain date. The upper division courses tend to enroll more mature students and the list of deadlines might not be as long. But in either case, failure to meet any deadline is factored into the project/paper grade, which counts for 15% of the course grade.

The evaluation of a student project is much more than making a judgment on the quality of the student's writing. The student typically meets with the instructor two or three times during the last month of the course. During these meetings the instructor can evaluate the student's preparation, judge how much effort is being put into the project and to the course in general, and when appropriate gently push a student to work harder. The project assessment process also helps with the overall assessment of the student.

When the project finally is submitted, the students also must include the related research articles. The instructor then goes through the research articles again before actually reading the paper. The projects are evaluated both on grammar and structural flow and on how well the students understand and discuss the new statistical methods used in the research articles. For example, a research article might use a repeated measures design in which patients are randomized into treatment and control groups, with the results measured over a period of time. Then the student writing the paper must address the fact that the repeated measurements are correlated and that standard statistical techniques do not work. This must be followed by a description and evaluation of the more sophisticated methods that this particular setting requires.

3.1.7 Conclusion

Many statistical educators feel that providing cookbook courses in statistics only furthers the misconception that statistical methodology is a static domain with only limited applicability in practice. In contrast, the above examples show that the field of applied statistical research is constantly evolving to meet the needs of the end-user. Applied statisticians who engage in consulting are well aware that it is not enough to possess a statistical toolkit from which the proper tool is produced to solve the researcher's problem. The successful statistical consultant must develop the ability to meet and address challenging statistical problems with innovative, novel solutions.

As has been seen throughout this article, class projects are beneficial in this regard. They also are useful in underscoring the dynamic nature of our domain by highlighting the fact that *statistical consultants are continuously learning new techniques and refining old tools* so as to better address the problems of researchers. Class projects remind students never to simply accept the results of a given statistical test or prediction without first understanding the underlying assumptions and limitations. Students learn this as they observe situations in which medical researchers inappropriately apply statistical techniques which violate necessary model assumptions. They also learn this by being exposed to situations in which the preliminary (incorrect) analysis leads researchers to an incorrect conclusion. These projects help students develop the critical eye and questioning spirit required of a successful researcher seeking to better understand their field. Finally, by requiring students to go independently beyond the standard course content, the class projects provide students with the sense of confidence needed to master new fields of applied statistics and to become intelligent and critical statistical consumers and consultants.

References

1. A. Agresti, *An Introduction to Categorical Data Analysis*, Wiley, New York, 1996.
2. P. Armitage and T. Colton, eds., *Encyclopedia of Biostatistics* (6 volumes), Wiley, New York, 1998.
3. A.C. Atkinson and A.N. Donev, *Optimum Experimental Designs*, Clarendon Press, Oxford, 1992.
4. D.M. Bates and D.G. Watts, *Nonlinear Regression Analysis and its Applications*, Wiley, New York, 1988.
5. C.R. Boyle, "A Problem-Based Learning Approach to Teaching Biostatistics," *Journal of Statistics Education*, 7 (1999) (obtained on the Web at the website http://www.amstat.org/publications/jse/secure/v7n1/boyle.cfm.
6. B.L. Chance and A.J. Rossman, "Sequencing Topics in Introductory Statistics: A Debate of What to Teach When," *The American Statistician*, 55 (2001) 140–144.
7. M. Davidian and D.M. Giltinan, *Nonlinear Models for Repeated Measurement Data*, Chapman & Hall, New York, 1995.
8. E. Downing, V.V. Fedorov and S. Leonov, "Extracting Information from the Variance Function: Optimal Design," in *mODa6 - Advances in Model-Oriented Design and Analysis*, Physica-Verlag, 2001.
9. W.J. Ewens and G.R. Grant, *Statistical Methods in Bioinformatics*, Springer, New York, 2001.
10. T.M. Gerig, U. Blum, and K. Meier, "Statistical Analysis of the Joint Inhibitory Action of Similar Compounds," *Journal of Chemical Ecology*, 15 (1989) 2403–2412.
11. L.M. Haines, G.P.Y. Clarke, E. Gouws and W.F. Rosenberger, "Optimal Design for the Testing of Anti-malarial Drugs," in *mODa6 - Advances in Model-Oriented Design and Analysis*, Physica-Verlag, 2001.
12. L.M. Haines, T.E. O'Brien, and G.P.Y. Clarke, "Kurtosis and Curvature Measures for Nonlinear Regression Models," submitted to *Statistica Sinica* (2002).
13. F.E. Harrell Jr., *Regression Modeling Strategies: With Applications to Linear Models, Logistic Regression, and Survival Analysis*, Springer, New York, 2001.
14. D.W. Johnson, and R.A. Wichern, *Applied Multivariate Statistical Analysis* (5th Edition), Prentice-Hall, New York, 2002.
15. W.N. Kernan, K.M. Viscoli, L.M. Brass, J.P. Broderick, T. Brott, E. Feldmann, L.B. Morgenstern, J.L. Wilterdink, and R.I. Horwitz, "Phenylpropanolamine and the Risk of Hemorrhagic Stroke," *New England Journal of Medicine*, 343 (2000) 1826–1832.
16. L. Krabbendam, M. Mayke, M.A. Derix, A. Honig, E. Vuurman, R. Havermans, J.T. Wilmink, and J. Jolles, "Cognitive Performance in Relation to MRI Temporal Lobe Volume in Schizophrenic Patients and Healthy Control Subjects," *Journal of Neuropsychiatry and Clinical Neurosciences*, 12 (2000) 251–256.
17. S. Lau, S. Illi, C. Sommerfeld, B. Niggemann, R. Bergmann, E. von Mutius, U. Wahn, and the Multicentre Allergy Study Group, "Early Exposure to House-dust Mite and Cat Allergens and Development of Childhood Asthma: a Cohort Study," *The Lancet*, 356 (2000) 1392–97.

18. R.C. Littell, G.A. Milliken, W.W. Stroup and R.D. Wolfinger, *SAS® System for Mixed Models*, SAS Institute Inc., Cary, NC, 1996.
19. S.G. Machado and G.A. Robinson, "A Direct, General Approach Based on Isobolograms for Assessing the Joint Action of Drugs in Pre-clinical Experiments," *Statistics in Medicine*, 13 (1994) 2289–2309.
20. P. McCullagh and J.A. Nelder, *Generalized Linear Models* (2nd Edition), Chapman & Hall, New York, 1989.
21. D. Moore, "New Pedagogy and New Content: The Case of Statistics," *International Statistical Review*, 65 (1997) 123–127.
22. J.C. Pinheiro and D.M. Bates, *Mixed-Effects Models in S and S-Plus*, Springer, New York, 2000.
23. C.A. Pope, R.T. Burnett, M.J. Thun, E.E. Calle, D. Krewski, K. Ito and G.D. Thurston, "Lung Cancer, Cardiopulmonary Mortality, and Long-term Exposure to Fine Particulate Air Pollution," *Journal of the American Medical Association*, 287 (2002) 1132–1141.
24. M.L. Samuels and J.A. Witmer, *Statistics for the Life Sciences* (3rd Edition), Prentice-Hall, New York, 2003.
25. M.E. Stokes, C.S. Davis and G.G. Koch, *Categorical Data Analysis Using the SAS® System* (2nd Edition), SAS Institute Inc., Cary, NC, 2000.
26. W.N. Venables and B.D. Ripley, *Modern Applied Statistics with S-Plus* (3rd Edition), Springer, New York, 1999.
27. S. Walrand, C. Chambon-Savanovitch, C. Felgines, J. Chassagne, F. Raul, B. Normand, M-C. Farges, B. Beaufrère, M-P. Vasson and L. Cynober, "Aging: A Barrier to Renutrition? Nutritional and Immunologic Evidence in Rats," *American Journal of Clinical Nutrition*, 72 (2000) 816–824.
28. J.H. Zar, *Biostatistical Analysis*, Prentice-Hall, New York, 1998.

Brief Biographical Sketch

Tim O'Brien is an Associate Professor of Statistics in the Department of Mathematics & Statistics at Loyola University Chicago. He is currently engaged in research in statistical education, bioinformatics and statistical modelling and optimal design for generalized nonlinear models with applications in biostatistics. He teaches an array of undergraduate and graduate courses in applied statistics and biostatistics.

3.2

Mathematical Biology Taught to a Mixed Audience at the Sophomore Level

Janet Andersen
Hope College

3.2.1 Introduction

The interface of mathematics and biology is an exciting area of research and an opportunity for changes in the curriculum. Funding agencies such as the National Science Foundation and the National Institute for Health are exploring or initiating programs to support this interface. *Biology 2010*, a report published by the National Research Council, delineates a curriculum for biology majors that is more mathematically intensive than what currently exists at most institutions. Talks related to mathematical biology can now be found at almost all national and regional MAA meetings.

At this point in time, most mathematical biology courses are either modelling courses designed for upper-level mathematics majors or lower level courses (typically with minimal mathematics prerequisites) designed for biology majors. Examples of textbooks used for such courses include *Mathematical Models in Biology* by Leah Edelstein-Keshet, *Mathematical Biology* by J.D. Murray, *Mathematical Models in Population Biology and Epidemiology* by Fred Brauer and Carlos Castillo-Chávez, *Understanding Nonlinear Dynamics* by Daniel Kaplan and Leon Glass, *Population Biology* by Alan Hastings, and *A Course in Mathematical Modeling* by Douglas Mooney and Randall Swift.

At Hope College, we chose to take a different approach. With the support of a National Science Foundation grant (NSF-DUE 0089021), we developed a team-taught mathematical biology course targeted at a mixed audience of mathematics and biology majors. The prerequisite for the mathematics students is completion of a linear algebra and differential equations course while the prerequisite for the biology students is completion of a sophomore-level course on ecology and evolutionary biology plus first semester calculus. We are assuming that the mathematics students do not know the biology and that the biology students do not know the mathematics. We organize the course by teaming these two audiences with the goal of critically reading biology research papers that incorporate mathematical models involving matrix analysis or ordinary differential equations.

3.2.2 Logistics and Course Objectives

Dr. Greg Murray, an ecologist at Hope College, and I developed this course. From the beginning, we decided to use biology research papers rather than a textbook. We want students to understand that the use of mathematical models is wide-spread in current biological research and that understanding such models is crucial to interpreting results. The course meets for three hours each week in lecture and for three hours each week in a wet lab. We think it is important that mathematics students gain an intuitive understanding of the biology by actually doing labs. Similarly, it is important that the biology students gain an intuitive understanding by doing the mathematical calculations (rather than solely using software packages).

Our course goals are for students to learn:

- How to communicate with someone in another discipline
- How to critically read research papers
- About areas of research that combine the study of mathematics and biology

These goals are intentionally process oriented rather than content oriented. Students do learn new mathematics and new biology. On the end of the course evaluation, biology students reported their learning of new mathematics was at 4.2 while the learning of new biology was 1.8 (both out of a scale from one to five). Mathematics students reported their learning of new biology was at 3.4 while the learning of new mathematics was 2.2 (both out of a scale from one to five). For both audiences, the item that received the highest rating was learning how to critically read research papers (with an average of 4.4 out of 5).

About half of the class time is devoted to lectures and class discussions on the biology and mathematics contained in the research papers. The remaining half is spent with students working together in interdisciplinary pairs. In these groups, students discuss the section of the research paper assigned for that day, comparing questions and vocabulary. They also work on mathematical worksheets that explain the computations used in the papers, such as finding eigenvectors or solving systems of first order linear differential equations.

Our first research paper is "A Stage-Based Population Model for Loggerhead Sea Turtle and Implications for Conservation" by D. Crouse, L. Crowder, and H. Caswell. This paper uses a stage-based matrix to describe the life-cycle of the loggerhead sea turtle. The biology topics include the complexities of conservation biology, the difficulty of obtaining reliable data, and the choices that must be made in the uses of resources. The mathematical topics include the use of matrix models, determining eigenvectors and eigenvalues, geometric series, and the derivation of the formulas for calculating sensitivities and elasticities. (Note: Sensitivities and elasticities are associated with the partial derivative of the dominant eigenvalue with respect to the entries of the matrix.) Students work together through the details of the paper, critiquing both the mathematics and the biology. While all mathematical calculations are initially done by hand, we use MAPLE for the more complicated computations. The lab that accompanies this paper is an extensive one that looks at the life cycle of flour beetles. We also do some work at the biology field station exploring the life cycle of garlic mustard.

The second research paper is "Mathematical Analysis of HIV-I Dynamics in Vivo" by A. Perelson and P. Nelson. This paper uses systems of primarily first order differential equations to study the dynamics of an AIDS infection and the impact of drug therapies. The biology topics include the physiology of the AIDS virus and its interactions with the immune system within a single host. The mathematical topics include phase portraits, equilibrium solutions, finding solutions to systems of linear first order differential equations using eigenvectors and eigenvalues, and basic probability. The accompanying molecular biology lab allows students to study e-coli bacteria infected with the lambda phage. They use Polymerase Chain Reaction (PCR) techniques to amplify the viral DNA and a spectrophotometer to quantify the rate of the infection.

Grades are based on questions from the research papers, quizzes on both the mathematical and biological concepts, lab reports, and oral presentations. Students do group oral presentations on the research papers covered in class. These presentations are each 15-20 minutes, typically done in PowerPoint, and must explain the connections between the biology and the mathematics. Students are required to find additional sources to supplement the information found in the paper.

With the Crouse, et al., paper on Loggerhead Sea Turtles, each group was assigned a section of the paper. The first group was responsible for background on sea turtle biology and age-classified population projection models. They developed a life cycle diagram for the turtles, showed how to translate it into matrix form, and discussed how to determine lambda, the stable age distribution and the reproductive value. Since the data used in the Crouse paper was obtained from other sources, this group was responsible for finding those sources (referenced in the Crouse paper) and explaining how the data was obtained and modified. In particular, they had to highlight the assumptions that were inherent in the data. The second group was responsible for explaining how to reformulate the age-classified model into a stage-classified model as well as the meaning and derivation of the formulas involving sensitivities. This group supplemented the paper by proving the formulas that we had stated (but not proved) in class. One of the references they used was *Matrix Population Models* by Hal Caswell. The third group presented the results of the analysis done in the paper and discussed the biological interpretation. They supplemented the paper by writing MAPLE code to reproduce the analysis and using websites to research the current status of the loggerhead sea turtles. Each group gave a 20-minute presentation, with an additional five minutes for questions. Each student was required to ask at least one question and a small part of the grade reflected the presenter's ability to answer these questions. Prof. Murray and I met at least once with each of the groups to critique their talk and slides. This critique was crucial in improving the quality of the presentations.

A significant part of the grade is based on the final presentation of a self-selected research paper. This must be a biology research paper that incorporates both data and a mathematical model. Examples of papers used this year include population dynamics of cheetahs, predator-prey spatial dynamics exhibited by spider mites, deforestation caused by the Mountain Pine beetle, and why drug therapies are not curing AIDS. These presentations are assigned two to three weeks before the end of the semester and the remaining classes are devoted to selecting a paper, working through the details, and creating the presentation. Each presentation is 30 minutes and both biology and mathematics faculty are invited. The final project allows us to assess the students' ability to read a biology research paper independently and to work through the details of both the mathematics and the biology. Students find that the mathematical models carry hidden assumptions that are ignored if one skips the mathematics. They also critique whether the mathematical model used appears to be the most appropriate for the biological situation. We have been quite pleased with the quality of the presentations.

3.2.3 Assessment and Results

We have now taught the mathematical biology course twice, once in spring 2002 and again in spring 2003. Both times we were able to have about 60% of the students from biology and about 40% from mathematics. Students consistently rate the course highly. Students appreciate that the course focuses on the interplay between the disciplines, uses research papers as the primary text, and uses oral presentations rather than exams as the primary assessment. In response to the question "I am glad I took the course because," student comments included:

> It was good to be challenged.
> I enjoyed working through research papers.
> I learned a lot about the interactions between math and biology.
> That I heard another discipline's point of view and learned how helpful math is.
> It really helped me read research papers critically.

My presentation skills improved, I learned how to read a paper and not gloss over the math and interacted with students and prof from other disciplines that taught me a lot and were <u>fun</u> to work with.

It is what you look forward to in math. When what you know actually applies to real life.

Student responses to the question "I am most frustrated with the course because" focused on the unevenness of the workload (primarily from the students who took the course in 2002), the difficulties associated with the logistics of a team-taught course, and frustration that some biology research papers are not necessarily using mathematical models appropriately (responding to papers used for the final presentation).

With the support of the NSF grant, we were able to hire summer research students to help develop materials and labs for the course. These students have always been one biology major and one mathematics major and, where possible, students who have taken the course. Taking the course and doing research in the summer has resulted in mathematics majors taking biology courses and pursuing careers in mathematical biology. In fact, two of our mathematics majors started graduate work in mathematical biology in the fall of 2003.

One of the most exciting and unanticipated outcomes from the course is the increased interaction between the biology and mathematics departments at Hope College. In addition to collaborating on the course, Dr. Murray and I have begun to collaborate on his research. The chair of the biology department, Dr. Tom Bultman, recently submitted a biology research grant that included summer stipends for both a mathematics student and a mathematics faculty member. Dr. Leah Chase, a neuroscientist at Hope College, and I meet weekly to discuss ways to use mathematics in her research and she team-taught the mathematical biology course with me in spring 2004. Collaborations on curriculum have spread to collaborations on research which, in turn, have informed the collaborations on curriculum.

3.2.4 Implementation Issues

One of the most difficult issues in implementing such a course is finding a collaborator and convincing the administration that a team-taught course is worthwhile. Finding a collaborator means getting to know the faculty in the biology department and gravitating towards those that appreciate the use of mathematical models. Ecologists are a natural place to start. Getting the administration to approve a team-taught course is often more difficult. In our case, we are able to do this since the course has a three hour lecture and a three hour lab. I get credit for the lecture and the biologist gets credit for the lab (which counts as a course). This way, we each receive credit for one course. We chose to structure it, however, so that both of us are always present for both the class and the lab.

Once a collaborator is found and a team-taught course has been approved, there remains the difficulty of finding the students. We cross-list this course so that a student may receive either biology credit or mathematics credit for the course. In both cases, it counts towards the major. If a mathematics student chooses to take the course for biology credit, this will count towards the general education requirement of taking a lab science. If a biology student chooses to take the course for math credit, it counts towards their mathematics cognate requirement. We have found that having the course satisfy multiple requirements is essential to attracting an audience. Secondly, I cannot overemphasize the importance of advertising the course. This includes putting up flyers, sending e-mail, talking to advisors, and talking individually to students. Most students are unaware of the growing opportunities for people able to work at the interface of mathematics and biology. Most biology students are unaware of the increasing use of mathematics in biological research. It takes time and effort to convince both audiences that a course in mathematical biology is useful, especially since this includes convincing mathematics students to take a course that meets six hours a week! However, there continues to be a subpopulation of mathematics majors who are looking for opportunities to apply mathematics as well as biology majors who recognize the unique benefits found in taking such a course, particularly as preparation for graduate school. Our best recruiting device has been the students who have taken the course or worked with us over the summer. Many students find the focus of the course on learning how to critically read research papers, do effective oral presentations, and communicate with those from a different disciplinary perspective to be appealing and a welcome change from traditional course work.

3.2.5 Conclusion

I am convinced that training mathematics majors to apply mathematics to other disciplines is an important skill that is often ignored or trivialized in traditional curricula. In particular, the benefits derived from pairing mathematics majors with students from another discipline to solve non-trivial, real problems is tremendous. Although the approach we have taken is to pair mathematics and biology, I believe this would work equally well in pairing other disciplines such as mathematics and physics or chemistry or one of the social sciences. However, there are some elements that I think are essential:

- Both groups of students (and faculty) must clearly understand that they cannot solve the problem independently but rather that it requires expertise from both disciplines.
- Students (and faculty) must get beyond a fear of asking stupid questions. Both groups are novices and both groups are experts.
- It is crucial that there be a hands-on component for both disciplines so that both groups gain an intuitive understanding of what it means to work in another discipline.
- The fact that different disciplines define problems differently and apply different strategies to reach a solution must be seen as an asset rather than as a barrier.
- Students (and faculty) must realize that interdisciplinary communication is difficult and takes time. The differences in vocabulary and perspective make it very easy to talk to one another and yet not communicate.

Creating this course, team-teaching with faculty from other departments, and collaborating with others on research projects has been one of my most rewarding and exciting experiences. It has allowed me to capitalize on a love of learning that I value in myself and try to foster in my students. It has been an opportunity to integrate research and education for both the faculty and the students. And it has provided our students with an example of the interdisciplinary team approach that closely resembles the environment in which many mathematicians work in the real world.

References

1. F. Brauer and C. Castillo-Chávez, *Mathematical Models in Population Biology and Epidemiology*, Springer, New York, 2000.
2. H. Caswell, *Matrix Population Models, 2nd Edition*, Sinauer Associates, Inc. Publishers, Sunderland, Massachusetts, 2001.
3. D. Crouse, L. Crowder, H. Caswell, "A Stage-based Population Model for Loggerhead Sea Turtles and Implications for Conservation," *Ecology*, 68 (1987)1412–1423.
4. L. Edelstein-Keshet, *Mathematical Models in Biology*, McGraw-Hill, Boston, 1988.
5. A. Hastings, *Population Biology*, Springer, New York, 1997.
6. D. Kaplan and L. Glass, *Understanding Nonlinear Dynamics*, Springer, New York, 1995.
7. D. Mooney and R. Swift, *A Course in Mathematical Modeling*, The Mathematical Association of America, 1999.

8. J.D. Murray, *Mathematical Biology*, Springer-Verlag, New York, 1989.
9. A. Perelson and P. Nelson, "Mathematical Analysis of HIV-I Dynamics in Vivo," *SIAM Review*, Vol. 41, No. 1 (1999) 3–44.
10. P. Whitacre, editor, *Bio 2010: Transforming Undergraduate Education for Future Research Biologists*, The National Academies Press, Washington, DC, 2003.

Brief Biographical Sketch

Janet Andersen taught high school in east Texas before receiving her PhD from the University of Minnesota. She has been the Principal Investigator on two previous NSF grants. The first grant resulted in materials for an innovative precalculus course and the second grant resulted in materials based on the use of mathematics in public media for a general education course. She is currently an Associate Professor, Chair of the Mathematics Department at Hope College, and the Director of the Pew Midstates Science and Mathematics Consortium.

3.3

A Geometric Approach to Voting Theory for Mathematics Majors

Tommy Ratliff
Wheaton College

3.3.1 Introduction

In the Spring of 2002, I taught an upper level course in Game and Voting Theory at Wheaton College, a small liberal arts college, in Norton, Massachusetts. There were twelve students enrolled in the course. Most were majors in Mathematics, Mathematics/Computer Science, or Mathematics/Economics, although there was one History major and one Psychology major. All the students had taken our Discrete Mathematics course, which serves as our introduction to proofs class. We spent the first half the semester on game theory and the remaining seven weeks on voting theory. In this paper, I will focus on the voting theory part of the course and give a brief tour through some of the course content. I also will describe the structure of the assignments and directions that I would like to take the course in the future.

Approximately four years ago, I changed my research area to voting theory from algebraic topology, in part because the questions and answers often are accessible to undergraduates, even if the proofs are not. The course provided an opportunity to expose the students to an active area of research and to explain recent results. For many students, this was the first time they had seen theorems from the last half of the 20th century, and it was certainly the first time any of the students had seen results published in the 21st century. This made quite an impression on several of the students, who commented on this in their course evaluations.

The students' primary context for understanding problems with voting came from the 2000 Presidential election, where they identified dangling chads, the Electoral College, and the Supreme Court as the main issues. Most of them had not realized that there can be problems in the underlying voting methods, even if there is perfect information about the voters' preferences. The students were surprised to learn that when there are more than two alternatives, the outcome can depend on the procedure used as much as on the voters' preferences and they were pleasantly surprised that significant mathematical analysis can help in understanding why these inconsistencies occur. For most of the students, this was also the first time they had seen non-statistical mathematical methods applied outside of the sciences.

There are many materials that cover voting theory designed for lower level courses for non-majors, but I wanted to give the students an introduction to the geometric framework that underlies some of the very interesting recent results and is also the basis for my own research. Unfortunately, there is no text that addresses this material. The students used Don Saari's *Chaotic Elections! A Mathematician Looks at Voting* [3] as a reference, and I used Hannu Nurmi's *Voting Paradoxes and How to Deal with Them* [1] as an additional source. Not having a text was one of the weaknesses of the course, since the students did not have a text to reference for their problem sets. In most courses, I like to emphasize that students read the text before class so that they have some familiarity with the material beforehand, but this was not an option in this course.

I chose to focus the course on paradoxes and inconsistencies that can happen and to emphasize that these situations quite often occur not through any Machiavellian manipulation but rather happen inadvertently when people do not understand the implications of the voting method. I also wanted the students to think of voting in a larger context, not just elections for public office, but also votes by committees, a group deciding where to eat dinner, or the selection of the figure skating medalists at the Winter Olympics. There were two fundamental questions that came up repeatedly during the semester:

1. Given the specific preferences of all of the voters, which outcomes are possible by varying the voting procedure?

2. Given a set of potential outcomes and corresponding voting procedures, is there a set of voters' preferences that can give these outcomes using the corresponding procedure?

The students seemed genuinely surprised by how badly things could go. Toward the end of the semester, one of the students came to ask about the way that her housemates for the coming year had determined who would get the single bedrooms in the house. She said that after taking the course she felt that something was significantly wrong with the method they had used... and she was correct. She told me that she would have never suspected this before taking the course. This is one consequence I was hoping for: I wanted the students to develop intuition about why paradoxes can occur with the intent of making them more informed consumers of choice procedures in their daily life.

3.3.2 A Brief Survey of the Course Content

The majority of the course focused on elections involving three candidates. If the full semester were spent on voting theory, then we would have gotten more deeply into elections with more candidates. Following the standard approach, we assume that all voters have complete, strict, transitive preferences among all possible outcomes. A listing of the voters' preferences is called a *profile*. I began the voting theory part of the course with the following (very hypothetical) example.

Example 1 *Wheaton just received a $50 million gift. A committee of students, faculty, staff, and administrators is formed to choose what to do. The three competing options are:*

A - Build a new Science Center
B - Massively renovate first-year and sophomore dorms
C - Reduce the comprehensive fee substantially
There are 35 people on the committee, and their preferences are:

$$
\begin{array}{llll}
10 & A > B > C & 6 & C > B > A \\
2 & A > C > B & 4 & B > C > A \\
7 & C > A > B & 6 & B > A > C
\end{array}
$$

The fundamental question is: Given this profile, how do we aggregate the individuals' preferences to find the group preference? In the course, there was a standard catalog of voting methods that we studied.

1. **Plurality:** The voters indicate their top choice. In this case, the plurality outcome is $C > A > B$ with tallies $13 > 12 > 10$. Notice that plurality corresponds to the voters giving 1 point to their top-ranked candidate and 0 points to their second and third ranked. There are the normal variations where if no candidate receives a majority, then a runoff is held between the top two candidates, or where the bottom candidate is dropped (these variations are the same with three candidates).

2. **Antiplurality:** The voters give 1 point to their first and second ranked candidate and 0 points to their third. Whereas plurality emphasizes a voter's strong approval (their top-ranked candidate), antiplurality emphasizes their strong disapproval (their bottom-ranked candidate). In this example, the antiplurality outcome is $B > A > C$ with tallies $26 > 25 > 19$.

3. **Borda Count:** The voters give 2 points to their top-ranked candidate, 1 point to their second-ranked candidate, and 0 points to their last ranked candidate. The outcome for this example is $A > B > C$ with tallies $37 > 36 > 32$.

4. **Other Positional Methods:** Plurality, antiplurality, and the Borda count are specific instances of *positional voting methods* where weights $w_1 : w_2 : w_3$ are assigned to a voter's first place, second place, and third place candidates, respectively. Plurality has weights $1 : 0 : 0$, antiplurality has weights $1 : 1 : 0$, and the Borda count has weights $2 : 1 : 0$.

    If we used weights $4 : 1 : 0$, then the outcome is $A > C > B$ with tallies $61 > 58 > 56$. The important point to notice is that by using four different positional methods, we have obtained four different rankings for the same election! Depending upon the procedure chosen, the committee could reasonably pick any of the three alternatives.

5. **Condorcet Criterion:** If one candidate beats all the other candidates in head-to-head elections, then this candidate is called the *Condorcet winner*. In our example, A is the Condorcet winner since it beats B by a count of $19 > 16$ and beats C by a count of $18 > 17$. One of the fundamental problems in voting theory is that the Condorcet winner may not always exist since there can be a cycle among the candidates.

6. **Dodgson's Method:** This method was proposed by Charles Dodgson (aka Lewis Carroll) and attempts to find the candidate that is closest to being the Condorcet winner if the Condorcet winner does not exist. One of the peculiarities of Dodgson's Method is that there is a profile with five candidates where A is the Dodgson winner, but if every voter brings three friends with exactly the same preference, then B becomes the Dodgson winner. I gave the students this profile in a homework set and asked them to calculate the Dodgson winner, and they found this quite remarkable. This example appeared in a paper of mine in 2001 [2], although this general property of Dodgson's Method has been known for some time.

7. **Approval Voting:** In approval voting, each voter indicates which candidates they approve of, and each of these candidates receives 1 point. Some voters may vote for one candidate, while others may vote for two. We assume that no voter will select no candidates or all three since this would be equivalent to expressing no opinion. One approval voting scenario from our example is:

		Vote for one	Vote for two
10	$A > B > C$	5	5
2	$A > C > B$	1	1
7	$C > A > B$	6	1
6	$C > B > A$	6	0
4	$B > C > A$	1	3
6	$B > A > C$	6	0

In this case, the approval voting outcome is $C > B > A$ with tallies $17 > 15 > 13$. This is yet another ranking distinct from the four rankings we obtained from the positional methods. In fact, by modifying which voters select one candidate and which voters select two, it is possible to obtain all thirteen of the rankings on three candidates (including ties) from this profile using approval voting.

Example 1 was quite a disturbing scenario for the students; no voter changed their rankings, but depending upon the voting method used, the outcome could differ dramatically. With the coverage of the irregularities in Florida during the 2000 Presidential election, the students knew that an election could be sensitive to including or excluding a small number of voters, and some were aware that the outcome can depend upon the method used (e.g., Gore won the national popular vote but Bush won the electoral college). However, they were all surprised at just how variable the outcome could be even with perfect information about the voters' preferences. In order to understand why these procedures give different outcomes, we used the *representation triangle* developed by Saari.

The Representation Triangle

For any method that tallies points for each candidate (e.g., any positional method or approval voting), we can measure the proportion of support for each candidate and assign the proportions to a point in $\mathbb{R}^3$. For example, using the profile in Example 1, plurality defines the point $\left(\frac{12}{35}, \frac{10}{35}, \frac{13}{35}\right)$ where the coordinates indicate the support for A, B, and C, respectively. The points for the other methods are given in Table 3.3.1.

Method	Tally			Point in $\mathbb{R}^3$
	A	B	C	
Plurality	12	10	13	$\left(\frac{12}{35}, \frac{10}{35}, \frac{13}{35}\right)$
Antiplurality	25	26	19	$\left(\frac{25}{70}, \frac{26}{70}, \frac{19}{70}\right)$
Borda count	37	36	32	$\left(\frac{37}{105}, \frac{36}{105}, \frac{32}{105}\right)$
Weights 4:1:0	61	56	58	$\left(\frac{61}{175}, \frac{56}{175}, \frac{58}{175}\right)$
Approval voting	13	15	17	$\left(\frac{13}{45}, \frac{15}{45}, \frac{17}{45}\right)$

Table 3.3.1. Tallies for Example 1

An important observation is that a point (a, b, c) arising from any profile and any voting method will satisfy

$$a + b + c = 1 \quad \text{and} \quad a, b, c \geq 0.$$

Thus, all of these points must lie on the portion of the plane $a + b + c = 1$ in the first octant, as shown in Figure 3.3.1.

Notice that each vertex on the triangle corresponds to a unanimous plurality outcome. For example, the point $(1, 0, 0)$ corresponds to the plurality outcome for a profile where every voter has A top ranked. Therefore, the proximity of the point to the vertices indicates the ranking of the candidates as determined by the procedure in use. If we take an orthogonal view of the representation triangle as shown in Figure 3.3.2, we see that it is naturally broken into six regions, each of which defines a complete transitive ranking of the candidates.

3.3 A Geometric Approach to Voting Theory for Mathematics Majors

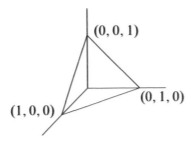

Figure 3.3.1. The image of the profiles in $\mathbb{R}^3$

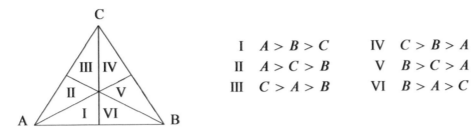

I	$A > B > C$		IV	$C > B > A$
II	$A > C > B$		V	$B > C > A$
III	$C > A > B$		VI	$B > A > C$

Figure 3.3.2. Regions in the representation triangle

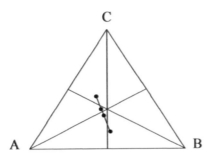

Figure 3.3.3. Procedure line for Example 1

Figure 3.3.3 shows the points corresponding to the four positional outcomes for Example 1. It may be somewhat surprising that these points appear to be colinear. The key observation is that any position method $w_1 : w_2 : w_3$ can be normalized so that it is of the form $1 : s : 0$ with $0 \leq s \leq 1$ by first subtracting w_3 from all the weights and then dividing by $w_1 - w_3$. Therefore, the outcome for any position method will lie on the line segment connecting the two extreme values for s. These extreme values are defined by $s = 0$ (plurality) and $s = 1$ (antiplurality), and we call the line segment the *procedure line* for the profile.

This observation allows us to easily identify *all* possible outcomes for a profile by using a positional method. For example, there is no positional method that will give an outcome of $B > C > A$ for Example 1. There are several results that are now immediate:

- A given profile can have at most four strict outcomes using positional methods since this is the maximum number of regions that a line segment can intersect.
- If plurality and antiplurality give the same ranking, then all positional methods give this same ranking on the profile.

Once we have identified the plurality and antiplurality outcomes, all possible positional outcomes for the profile are clear. There is some restriction on the coordinates for plurality and antiplurality. First notice that the antiplurality tally for each candidate is at least as large as the plurality tally and that the denominator for the antiplurality point is twice the denominator for the plurality point. Thus, the maximum proportion that a candidate can receive under antiplurality is $\frac{1}{2}$ since every voter selects two candidates, and the antiplurality proportion must be at least half of the plurality proportion. Therefore, if (p_1, p_2, p_3) is the plurality point in the representation triangle for a profile and (ap_1, ap_2, ap_3) is the antiplurality point, then

$$\frac{1}{2} p_i \leq ap_i \leq \frac{1}{2} \text{ for } i = 1, 2, 3$$

This relationship is important when trying to create examples where the plurality point lies in one region and the antiplurality point lies in another.

Decomposition of $\mathbb{R}^6$

The use of the procedure line and representation triangle describes the extreme outcomes that can be achieved by a given profile, but they do not really explain why different procedures give different outcomes on the same profile. For this, we use a decomposition of profiles into fundamental subspaces that was developed by Saari.

By picking a consistent ordering for the six rankings of the three candidates, we can easily identify a profile with a point in $\mathbb{R}^6$. For example, if we use the same ordering as in Figure 3.3.2, then Example 1 corresponds to the point $(10, 2, 7, 6, 4, 6)$. Now consider the six profiles shown in Table 3.3.2: B_A and B_B are the *Basic Profiles* for A and B, respectively; R_A and R_B are the *Reversal Profiles*; C is the *Condorcet Profile*; and K is the *Kernel Profile*.

$$K = (1, 1, 1, 1, 1, 1) \qquad C = (1, -1, 1, -1, 1, -1)$$
$$B_A = (1, 1, 0, -1, -1, 0) \qquad R_A = (1, 1, -2, 1, 1, -2)$$
$$B_B = (0, -1, -1, 0, 1, 1) \qquad R_B = (-2, 1, 1, -2, 1, 1)$$

Table 3.3.2. Our basis vectors for $\mathbb{R}^6$

These profiles identify natural symmetries that occur in the space of all profiles. One may object to the negative number of voters in all but K. However, the Kernel profile has essentially no impact on any voting method and allows us to add multiples of K to the other profiles to get a non-negative number of voters. The fundamental properties of these profiles, which the interested reader can verify, are:

- These form a basis for $\mathbb{R}^6$.
- K, R_A, R_B have no impact on pairwise outcomes or Borda count.
- All positional methods give the same outcome on B_A and B_B. In particular, B_A boosts A's count compared to B and C, and B_B boosts B's count compared to A and C.
- C explains all differences between positional methods and pairwise outcomes.
- R_A and R_B explain all differences among the positional methods, including the Borda count, plurality, and antiplurality. Depending on the specific weights of the method, R_A will either boost or diminish A compared to B and C, and R_B will either boost or diminish B compared to A and C.

This basis allows us to generate profiles that exhibit conflicts. For example, if we want to create a profile where the Borda count gives $B > C > A$, there is no Condorcet winner, and the outcome using weights $5 : 2 : 0$ is $A > B > C$, then we can use the basis profiles as our building blocks.

- Begin with $-B_A + B_B$ to give the Borda outcome of $B > C > A$. None of the other profiles we add will affect the Borda count.

- Add a strong Condorcet term ($5C$ will do) to give a cycle in the pairwise votes of $A > B$, $B > C$, $C > A$. The reversal terms we add next will not affect the Borda count or pairwise elections.
- In order to boost A over B and C to get the $A > B > C$ outcome using the weights $5 : 2 : 0$, a reversal term of $11 R_A$ is sufficient.
- This gives a profile which, unfortunately, has a negative number of voters. By adding $26K$, we wind up with the profile $-B_A + B_B + 5C + 11 R_A + 26 K$, or

41	$A > B > C$	33	$C > B > A$
30	$A > C > B$	44	$B > C > A$
8	$C > A > B$	0	$B > A > C$

which we can verify has the desired properties.

We can also use the basis to explain why a given profile has different outcomes for different procedures. For example, the profile from Example 1 has decomposition

$$\frac{5}{6} B_A + \frac{2}{3} B_B + \frac{7}{6} C - \frac{7}{6} R_A - \frac{5}{3} R_B + \frac{35}{6} K.$$

The $\frac{5}{6} B_A + \frac{2}{3} B_B$ component explains why the Borda count is $A > B > C$; the $\frac{7}{6} C$ term is not strong enough to prevent A from being the Condorcet winner; and the $-\frac{7}{6} R_A - \frac{5}{3} R_B$ term introduces the conflict with the Borda count, plurality, antiplurality, and the weighted system $4 : 1 : 0$.

Sample Problems

These are some of the problems I assigned on homework and take-home exams that are related to the procedure line and the decomposition of $\mathbb{R}^6$.

1. Create examples of profiles with three candidates that have the following properties or explain why no such profile exists. If the profile exists for each outcome, give the weights for a procedure that determines that outcome.

    (a) All positional methods give the ranking of $A > B > C$.

    (b) By varying the positional procedure, the profile gives precisely the following strict positional outcomes:

    $A > B > C \quad A > C > B \quad C > A > B$

    (c) Repeat part (b) but with outcomes

    $C > B > A \quad B > C > A \quad B > A > C \quad A > B > C$

    (d) Repeat part (b) but with outcomes

    $C > A > B \quad C > B > A \quad B > A > C$

    (e) The only possible outcomes using a positional procedure are

    $B \sim C > A \quad B > C > A \quad B > A \sim C$

    where $B \sim C$ indicates a tie between B and C.

    (f) There are (at least) five different strict transitive outcomes that can be obtained by using five different voting methods.

2. Let q_0 denote the plurality point in the representation triangle for a profile and q_1 denote the antiplurality point. In each case, create a profile with these outcomes or explain why it is impossible.

(a) $q_0 = (\frac{1}{4}, \frac{5}{12}, \frac{1}{3})$ $q_1 = (\frac{3}{8}, \frac{1}{4}, \frac{3}{8})$
(b) $q_0 = (\frac{1}{32}, \frac{2}{3}, \frac{31}{96})$ $q_1 = (\frac{1}{5}, \frac{8}{15}, \frac{4}{15})$
(c) $q_0 = (\frac{1}{3}, \frac{1}{5}, \frac{7}{15})$ $q_1 = (\frac{3}{10}, \frac{1}{3}, \frac{11}{30})$
(d) $q_0 = (\frac{3}{22}, \frac{5}{22}, \frac{7}{11})$ $q_1 = (\frac{3}{7}, \frac{3}{7}, \frac{1}{7})$

3. For each profile that you determined was possible in problem 2, give all possible outcomes for this profile (including ties) using a positional method. In each case, give the positional method that determines the outcome.

4. For each profile that you determined was possible in problem 2, give all possible outcomes for this profile (including ties) using approval voting. In each case, give an approval voting ballot that determines the outcome.

5. Create examples of profiles with three candidates that have the following properties or explain why no such profile exists.

   (a) The only strict positional outcomes are $A > C > B$ and $B > C > A$.
   (b) The plurality outcome is $A > B > C$, the antiplurality outcome is $C > A > B$, and the Borda count gives $B > C > A$.
   (c) The Borda count is $A > B > C$, the pairwise outcome is $B > A > C$, and the plurality outcome is $C > A > B$
   (d) There is no Condorcet winner, the outcome using weights $(5, 2, 0)$ is $A > B > C$ and the Borda count outcome is $B > C > A$.
   (e) The plurality outcome is $B > A > C$, the Borda count gives $C > A > B$ and A is the Condorcet winner.

Other Topics

There were several other topics in voting theory that we also touched on during the semester.

1. There is a natural linear transformation $\mathbb{R}^6 \to \mathbb{R}^3$ from the space of profiles on three candidates to a space of pairwise outcomes. The key point is that each of the eight octants in $\mathbb{R}^3$ corresponds to a ranking of the three candidates: six correspond to the transitive rankings, one corresponds to the cycle $A > B, B > C, C > A$, and the other to the reverse cycle $A > C, C > B, B > A$. Using this framework, the students were able to show that the Condorcet winner can never be ranked last by the Borda count.

2. The No-Show paradox shows that it can be advantageous for a large number of voters to choose not to participate. Specifically, there exists a profile with 100 voters where A is the winner under plurality with runoff, but if 47 voters with preference $B > C > A$ abstain, then C becomes the winner. In other words, 47% of the voters have motivation to not vote so that their second place candidate is elected rather than their last place candidate.

3. We briefly discussed strategic voting and manipulation where a voter may improve the outcome of the election from their perspective by misrepresenting their true preferences. For example, if A narrowly beats B by one point using the Borda count, then a voter with preference $B > A > C > D$ could make B the Borda count winner by using the strategic vote of $B > C > D > A$. By moving A down in their ranking, the voter has reduced A's tally by two points. The celebrated Gibbard Satterthwaite Theorem shows that every non-dictatorial voting procedure is subject to manipulation.

4. Arrow's Theorem is perhaps the best known result in voting theory. It shows that the only procedure that satisfies a particular set of fairness criteria is a dictatorship. There is some argument, however, about whether the criteria are too restrictive and do, in fact, force the extremely negative result.

3.3.3 Assignments and Future Plans

During the voting theory part of the semester, the students had an individual homework set, a group homework set, a take-home exam, a book review, a group presentation, and a comprehensive take-home final exam that was handed out during the last week of classes. Ideally, I would have preferred more homework assignments, but generating all of the homework and exam questions took longer than I had hoped.

An interesting consequence of the homework assignments was that the final grades were much higher than in other upper-level courses I have taught. I believe that this was a result of the nature of the assignments: the students knew when their profiles met the desired criteria. Their answers did not necessarily come quickly (on the course evaluations, ten out of twelve students indicated that they spent more time on this course than on their other courses), but they were able to verify that their answers were correct. This was a quite different experience from my Real Analysis class the previous semester where the students were not so certain about their ϵ-δ proofs.

There were two pieces of software that helped the students with some of the routine calculations for three candidate profiles on the homework and take-home exams . I wrote a Maple worksheet to draw the procedure line for a given profile, as in Figure 3.3.3. The students found this quite useful since identifying the regions touched by the procedure line is very sensitive to plotting the plurality and antiplurality points accurately. In addition, the students in a computer science course wrote a Java applet to calculate the plurality, Borda, and pairwise outcomes for any three candidate profile.

For the group presentations, the students worked in groups of two or three. They picked topics, researched them on their own, and gave 15 minute presentations to the class. Most of the groups picked a specific voting method that is currently in use and showed how different results could be obtained either by slightly modifying the method or through strategic voting. Some of the topics were the proportional representation system used in municipal elections in Cambridge, Massachusetts, the method (and proposed reforms) used in judging Olympic figure skating, and the 1999 American League MVP ballot (where the students, who were all Boston Red Sox fans, showed how Pedro Martinez could have won). Since the students had already given presentations earlier in the semester during the game theory part of the course, the presentations on voting theory went fairly well.

My motivation for the book reviews was that there are many very interesting books about mathematics and mathematicians (as opposed to textbooks) that our students never read because the books do not fit into a specific place in the curriculum. I viewed this assignment as part of the students' general mathematical education, so I did not stress that the book need be related to the content of the course. Some of the books read by the students were *The Man Who Knew Infinity*, *The Code Book*, *The Man Who Loved Only Numbers*, *Flatterland*, and *A Beautiful Mind*. I emphasized to the students that their paper should be a critique of the book and not simply a summary. In particular, they should address the mathematical content and the writing style as well as the overall organization of the book. Most of the papers were between five and eight pages and the students enjoyed doing the reviews. I was very happy with the assignment, and I have continued to give it in my other upper division courses. The book review and group presentations combined to count for a total of 20% of each student's final grade in the course.

In Fall 2003, I expanded the voting theory portion into a full semester sophomore-junior level course. This allows more time to delve into manipulation and profiles with more than three candidates and to cover additional topics such as power indices (which measure the power of voters within a voting system), proportional representation systems, and questions of apportionment. As part of our new general education curriculum at Wheaton, this course will be connected with a course from political science on Congress and the Legislative Process. The political science professor and I will give guest lectures in each other's course, and we are planning a joint project for the students from both classes, although we have not yet worked out the exact format for this. The major challenge in planning the course is the lack of an appropriate text. I currently am developing a larger catalog of examples so that I can distribute solutions to the students. I also have expanded the Java applets to handle profiles with more candidates and add extra features, such as giving the decomposition of a profile in terms of the basis vectors.

3.3.4 Conclusions

Overall, I was fairly pleased with the voting theory portion in the course, especially for the first time teaching this material. One part of the take-home final asked the students to write a four-page essay where they explained the big picture of the course in their own words. From these essays, I believe the course succeeded in making them more skeptical about the process used by many groups to make decisions. For me, it was very rewarding to share my area of research with undergraduates and I believe the students benefited from seeing current research in an area that will almost certainly affect their daily lives.

References

1. H. Nurmi, *Voting Paradoxes and How to Deal with Them*, Springer-Verlag, 1999.
2. T.C. Ratliff, "A Comparison of Dodgson's Method and Kemeny's Rule," *Social Choice and Welfare*, 18 (2001) 79–89.
3. D.G. Saari, *Chaotic Elections!: A Mathematician Looks at Voting*, AMS, 2001.

Brief Biographical Sketch

Tommy Ratliff received his PhD from Northwestern University and is currently an Associate Professor of Mathematics at Wheaton College in Norton, Massachusetts. His email address is tratliff@wheatoncollege.edu, and there is more information about his classes on his website at

`http://www3.wheatoncollege.edu/~tratliff/`

3.4

Integrating Combinatorics, Geometry, and Probability through the Shapley-Shubik Power Index

Matthew J. Haines and **Michael A. Jones**
Augsburg College *Montclair State University*

3.4.1 Introduction

It is our belief that students compartmentalize mathematical techniques to be used solely for a specific problem or narrow set of problems. Ideally, students would develop a toolbox of mathematical techniques to analyze *any* problem from a multitude of perspectives. Analyzing simple weighted-voting games helps students develop varied approaches to problem solving while demonstrating how to use different mathematical skills in nontrivial, relevant ways. Such an analysis necessitates the integration of mathematical topics, including combinatorics, geometry, and probability.

This article serves as a primer for instructors so that they may introduce simple weighted-voting games and the Shapley-Shubik power index in order to relate voting theory to various topics in the curriculum. To encourage implementation and adaptation of this material, we include many examples and exercises. For this reason, this article may be used for self study by independent study students. These materials have been developed in a handful of courses at Montclair State University from spring 1999 to the present, including a general education requirement course, an upper level applied combinatorics and graph theory course, a graduate level course in combinatorial mathematics, and two independent study courses. One student has used this article as a self-study guide as a precursor to computational voting theory. These different levels of use are a testament to the diversity of mathematics that can be used to analyze simple weighted-voting games through the Shapley-Shubik power index. In this paper, we provide specific guidelines and suggestions on how to use the material developed in this article for different courses.

Simple weighted-voting games require only basic notions of sets, addition, and inequalities. The Shapley-Shubik power index is defined in terms of permutations of voters. Although both ideas are easy to understand, simple weighted-voting games and their Shapley-Shubik power indices can be rigorously analyzed to lead to interesting mathematics. The following is a sample of the types of mathematics and skills that are needed to analyze simple weighted-voting games and their Shapley-Shubik power indices. This list also provides a rough outline of the order that these topics appear in this paper. The analysis of Shapley-Shubik power indices of discrete, simple weighted-voting games requires concrete ideas about domain and range, applications of logic, properties of symmetry, permutations, and the addition of real numbers. It also requires the calculation of the number of nonnegative integer solutions of equations and the solution of equations with inequality constraints. The analysis of continuous, simple weighted-voting games also uses equivalence relations and partitions, the geometric relationship between inequalities and half-planes. It uses the area of a region to determine the likelihood that certain outcomes occur. Properties of discrete and continuous games are connected by limits of combinatorial identities that converge to areas of partitions.

We are not the first to suggest using simple weighted-voting games in the undergraduate curriculum. There are some exemplary materials that introduce simple weighted-voting games and the Shapley-Shubik power index in the context of modeling political interactions (Lampert [12] and Straffin [18],[19]), suitable for discrete mathematics and modeling courses. For this reason, we do not stress the modeling component in this work, although it is a part of our class presentation of this material. We believe that our approach asks more mathematical questions, especially geometric questions, of the students, while retaining its relevance to political science and modeling. There are other geometric approaches to mathematical political science that are accessible to undergraduate students, including those by Saari [15],[16]. These texts do not focus on simple weighted-voting games, but on election procedures.

One benefit of introducing this material to students is that they can read research on the evolution and formation of political institutions. In particular, there have been accessible analyses of the power indices of simple weighted-voting games modeling the European Economic Community (Brams and Affuso [2]), the European Union (Hosli [11], Berg [1], and Nurmi and Meskanen [14]), the International Monetary Fund (Dreyer and Schotter [5]), and the United States' Electoral College (Mann and Shapley [13]). At the heart of these articles are mathematical phenomena related to the institution at hand.

3.4.2 Simple Weighted-Voting Games and the Shapley-Shubik Power Index

Stockholders of a company are often allowed to vote "for" or "against" a potential company policy at a shareholders' meeting. Committee members often vote "yes" or "no" to arrive at a joint decision. Jurors vote "guilty" or "not guilty." All of these situations can be modeled by simple weighted-voting games. In the case of jurors, their votes are treated the same way. However, at a shareholders' meeting, someone who owns more shares of stock of the company has her vote count more; indeed, the vote counts for as many shares of stock as the stockholder has. Simple weighted-voting games can model these diverse situations, where voters' votes may be weighted differently. However, simple weighted-voting games model only those elections where two outcomes are possible: "yes" and "no." A measure is passed if enough voters vote "yes."

Definition 1: A simple weighted-voting game is a set of n voters $\{v_1, v_2, \ldots, v_n\}$, where voter i's vote carries the weight w_i, and a quota, a value that if the sum of the yes voters' weights is greater than or equal to the quota q then a measure passes. Denote a simple weighted-voting game by $[q; w_1, w_2, \ldots, w_n]$.

Typically, simple weighted-voting games are restricted by the following properties: w_i is a nonnegative integer for every i and $q > \frac{\sum_{i=1}^{n} w_i}{2}$. When the weights are restricted to nonnegative integer values, then we will call these *discrete simple weighted-voting games*. Consider the motivational examples of a jury, a committee, or the shareholders of a company. Every member of a jury must vote guilty for a defendant in a criminal trial to be found guilty. If any of the 12 members of the jury vote not guilty, then the jury is a hung jury and no decision is reached. (If the jurors do not all agree on guilty or not guilty, then the defendant of a trial may be re-tried.) Since a juror's vote is indistinguishable from another juror's vote, this jury can be modeled by the following simple weighted-voting game: $[12; 1, 1, 1, 1, 1, 1, 1, 1, 1, 1, 1, 1]$.

3.4 Integrating Combinatorics, Geometry, and Probability through the Shapley-Shubik Power Index

Notice that all voters' votes have the same weight.

There is not a unique simple weighted-voting game to model the jury's voting process. Indeed, [78; 12, 11, 10, 9, 8, 7, 6, 5, 4, 3, 2, 1] also represents the trial by jury because all of the voters must agree in the affirmative to return a guilty charge and the sum of the weights of the voters is the quota, $q = 78$. Although all of the jurors' votes are weighted differently, it is clear that they must vote unanimously to arrive at a guilty verdict, as before.

Consider a committee with a chairperson and three other members where in order for the committee to pass a measure, the chairperson and at least two of the other members must vote yes. This can be represented by the simple weighted-voting game [4; 2, 1, 1, 1], where the vote of the chairperson has weight 2. Notice that a measure can be passed only if the quota of 4 is reached. This only can be met by having the chairperson and two other members of the committee vote yes(since $2 + 1 + 1 \geq 4$) or by having the chairperson and all three other members vote yes(since $2 + 1 + 1 + 1 > 4$). The simple weighted-voting game [7; 3, 2, 2, 2] also models the relationship between the committee members' votes. In both games, a measure can only be passed if v_1 votes yes.

Suppose that a company has three stockholders and that each share of stock grants the owner of the stock one vote. Assume that the three stockholders own 55, 30, and 15 shares and that a measure is passed if a majority of the votes are in favor of the measure. This can be represented by the simple weighted-voting game: [51; 55, 30, 15]. Although the numbers in this simple weighted-voting game seem natural, realize that [501; 550, 300, 150] also models the voting relationship between the shareholders. A measure passes if v_1 votes yes and fails to pass if v_1 votes no.

Exercise 1: A Hiring Committee consists of a Personnel Director, a Team Manager, and three team members. The committee agrees to hire an applicant if, at the minimum, the director and all three team members or the director, team manager, and two team members agree to hire the applicant. Construct a simple weighted-voting game to model this situation.

Exercise 2: The quota of a simple weighted-voting game is required to be greater than half of the sum of the weights of all of the voters. To see why this is the case, consider the simple weighted-voting game [50; 50, 30, 20] and the measure "Voter 1 is the supreme ruler of the world." Does this measure pass? What about the measure "Voter 1 is not the supreme ruler of the world?" Does this measure pass? Explain why it is necessary to restrict the value of the quota.

Exercise 3: Suppose that the stock of a company splits. That is, assume that every share of stock is now worth 2 shares of stock. To preserve the relationship between the shareholders, what must be done to the quota? Explain.

Simple weighted-voting games provide a mathematical means to model political interactions. However, the mathematical framework presents opportunities not only to model but formally to ask, and to answer, questions inspired by the political setting. The most pertinent question is: "Who has political power?" From the examples, it is clear that a voter is better off having his vote carry a larger weight. But how is this related to political power? Power indices use the simple weighted-voting game structure along with insight about how political processes work to quantify the political power of players in a simple weighted-voting game. Before introducing the ideas needed to define the Shapley-Shubik power index, we present some terminology that represents extreme cases of the relationships between the voters. In fact, these terms are represented by the examples above.

Definition 2: A voter in a simple weighted-voting game
- is a *dictator* if she can pass a measure by voting Yes, even if all other voters vote No,
- has *veto power* if she can defeat a measure by voting against it, even when all other voters support the measure, and

- is a *dummy voter* if the outcome of an election *never* depends on her vote.

Notice that the chairperson in the committee example has veto power while the first voter in the stockholder example is a dictator. Furthermore, the other voters in the stockholder example are dummy voters. The jury example demonstrates that more than one voter, indeed all voters, may have veto power. Power indices provide a way to determine which simple weighted-voting games model the same situations. For applications of modeling real world situations with simple weighted-voting games, see COMAP [4].

The Shapley-Shubik power index focuses on the order of yes votes and who casts the deciding, or pivotal, vote. The pivotal voter has the power for this sequence of votes. The Shapley-Shubik power index of a voter is the number of times that a voter is pivotal over all possible sequences, or permutations, of the order of voters.

Consider the simple weighted-voting game $[3; 2, 1, 1]$. For the permutation, $v_1 \, v_2 \, v_3$, the second voter is the pivotal voter since $w_1 < q$ and $w_1 + w_2 \geq q$. A convenient method to compute the Shapley-Shubik power index is to list all of the permutations of the voters and to circle the pivotal voter for each ordering. All permutations of the three voters are listed in Figure 3.4.1 and the pivotal voters of the game $[3; 2, 1, 1]$ are indicated. Merely counting the number of times that each voter is circled yields the Shapley-Shubik power index. For $[3; 2, 1, 1]$, the Shapley-Shubik power index is 4:1:1.

Permutations

$v_1 \, (v_2) \, v_3$	$w_1 < q$	$w_1 + w_2 \geq q$	
$v_1 \, (v_3) \, v_2$	$w_1 < q$	$w_1 + w_3 \geq q$	
$v_2 \, (v_1) \, v_3$	$w_2 < q$	$w_2 + w_1 \geq q$	
$v_2 \, v_3 \, (v_1)$	$w_2 < q$	$w_2 + w_3 < q$	$w_2 + w_3 + w_1 \geq q$
$v_3 \, (v_1) \, v_2$	$w_3 < q$	$w_3 + w_1 < q$	
$v_3 \, v_2 \, (v_1)$	$w_3 < q$	$w_3 + w_2 < q$	$w_3 + w_2 + w_1 \geq q$

Figure 3.4.1.

Exercise 4: The Three Stooges meet regularly to discuss career options. Since Moe is the most recognizable and, in some sense, most essential Stooge, he has veto power on all business decisions. However, Moe is *not* a dictator. For Moe to pass a measure, either Larry or Curly has to vote "yes" also. Represent this situation with a simple weighted-voting game.

Exercise 5: Define equivalent simple weighted-voting games to be games where the voters have the same Shapley-Shubik power index. Find two such 3-voter games.

Exercise 6: For the simple weighted-voting game, $[q; w_1, \ldots, w_n]$, explain why the following is true: If the Shapley-Shubik power index of voter i is greater than the Shapley-Shubik power index of voter j, then $w_i > w_j$.

Exercise 7: For the simple weighted-voting game, $[q; w_1, \ldots, w_n]$, explain why the following is false: If $w_i > w_j$, then the Shapley-Shubik power index of voter i is greater than the Shapley-Shubik power index of voter j.

Exercise 8: Prove that a voter is a dictator in an n-player simple weighted-voting game if and only if his Shapley-Shubik power index is $n!$.

Exercise 9: Prove that a voter is a dummy voter if and only if her Shapley-Shubik power index is zero.

3.4.3 Possible Shapley-Shubik Indices

The Shapley-Shubik power index can be described by a function that takes a simple weighted-voting game with n voters to an n-tuple with nonnegative integer terms that sum to $n!$. More rigorously, the Shapley-Shubik power index can be defined by the value function on coalitions, although this introduces ideas that are not germane to this development (see *e.g.*, Shapley and Shubik [17] for deriving the Shapley-Shubik power index from the Shapley value, commonly used in game theory.) From our examples, we know that the Shapley-Shubik power index also depends on the quota. Rather than describe the Shapley-Shubik power index by a function with domain of n-tuples of weights together with a quota, we define separate functions for the total weight (*i.e.*, the sum of all voters' weights) and the quota.

Definition 3: A Shapley-Shubik power index is an image point of a function from the set of simple weighted-voting games with total weight w and quota q *into* the set of nonnegative integer solutions of

$$s_1 + s_2 + \cdots + s_n = n!.$$

More mathematically, the Shapley-Shubik power index is given by $S_{q,w}$ where $S_{q,w} : D \to R$ with domain

$$D = \{(w_1, w_2, \ldots, w_n) \mid w_i \geq 0 \text{ and } w_i \in \mathbb{Z}, \text{ for all } i, \text{ and } \sum_{i=1}^{n} w_i = w\}$$

and range

$$R = \{(s_1, s_2, \ldots, s_n) \mid s_i \geq 0 \text{ and } s_i \in \mathbb{Z}, \text{ for all } i, \text{ and } \sum_{i=1}^{n} s_i = n!\}.$$

The definition and functional notation of the Shapley-Shubik power index naturally leads to purely mathematical questions that will also have implications on applications. First and foremost, we may want to know if, for fixed w and q, the Shapley-Shubik power index is a one-to-one and/or onto function. Realize that for a 3-voter simple weighted voting game the Shapley-Shubik power index maps into the set of nonnegative integer solutions of $s_1 + s_2 + s_3 = 6$; there are $\binom{8}{2} = 28$ possible image points. Which of these possible image points are in the range (for some w and q)?

Exercise 10: Determine the 28 *possible* image points for the Shapley-Shubik power index for 3-voter, simple weighted-voting games. Use a combinatorial argument to explain why there are 28 possible image points.

Exercise 11: Use a combinatorial argument to count how many possible image points there are for the Shapley-Shubik power index function when there are n voters.

Exercise 12: Prove that if $m{:}n{:}p$ is a Shapley-Shubik power index for a simple weighted-voting game, then all permutations of m, n, and p are possible power indices, too.

Exercise 13: Explain why a voter is a dictator if she is ever pivotal in a permutation where she appears in the first position.

The following examples display the type of analysis necessary to determine the range of the Shapley-Shubik power index.

Example 1: 3:2:1 is not a possible Shapley-Shubik power index.

Consider the orderings of the voters in Figure 3.4.1. Since voter 1 is not a dictator, then he cannot be the pivotal voter in the sequences $v_1 \, v_2 \, v_3$ and $v_1 \, v_3 \, v_2$ (see Exercise 3). Hence, voter 1 must be pivotal in three of the four orderings:

$$v_2 \, v_1 \, v_3 \qquad v_2 \, v_3 \, v_1 \qquad v_3 \, v_1 \, v_2 \qquad v_3 \, v_2 \, v_1.$$

More specifically, voter 1 must be pivotal in $v_2\ v_3\ v_1$ or $v_3\ v_2\ v_1$. Voter 1 being pivotal in either sequence implies that $w_2 + w_3 < q$. So, voter 1 must be pivotal in both of these orderings. Therefore, voter 1 must be pivotal in one of $v_2\ v_1\ v_3$ and $v_3\ v_1\ v_2$. By Exercise 2, it follows that $w_1 > w_2 > w_3$ and $w_2 + w_1 > w_1 + w_3$. Voter 1 must be pivotal in the ordering $v_2\ v_1\ v_3$.

Voter 2's power index is two. Hence, voter 2 must be pivotal in two of the three orderings:

$$v_1\ v_2\ v_3 \qquad v_1\ v_3\ v_2 \qquad v_3\ v_1\ v_2.$$

One of these must be $v_1\ v_2\ v_3$. This follows since voter 1 is pivotal in $v_2\ v_1\ v_3$, indicating that $w_2 + w_1 \geq q$. However, if voter 2 is pivotal in either $v_1\ v_3\ v_2$ or $v_3\ v_1\ v_2$, then voter 2 must be pivotal in both. Hence, a contradiction on voter 2's power index being 2. So, 3:2:1 is not a possible Shapley-Shubik power index.

Example 2: 3:3:0 is a possible Shapley-Shubik power index.

Since voter 3 is a dummy voter, he can never influence the outcome of an election. Since neither voter 1 nor voter 2 is a dictator, both voters must agree in the affirmative and vote "yes" for a measure to pass. In the sequence of "yes" votes, which ever of voter 1 or voter 2 appears later in the sequence is the pivotal voter. Both voter 1 and voter 2 are equally likely to appear later in the six orderings. Hence, both are pivotal voters in three of the six orderings. The simple weighted-voting game [4; 2, 2, 1] has Shapley-Shubik power index 3:3:0.

Exercise 14: Realize that 2:2:2 is a valid Shapley-Shubik power index for certain values of w and q. However, the power index can be achieved by different relationships between the voters; one such way is unanimity rule while the other is majority rule. Explain.

Exercise 15: Explain why 5:1:0 is not a valid Shapley-Shubik power index.

Exercise 16: Can you ever get 2:2:2 for a Shapley-Shubik power index when $q = \frac{2}{3}w$?

Exercise 17: Determine which of the 28 points are images of the Shapley-Shubik power index for some 3-voter, simple weighted-voting game.

The analysis used in the previous two examples can be extended to any number of voters, albeit with some difficulty. However, thinking of which nonnegative integer solutions of $s_1 + s_2 + \cdots + s_n = n!$ are valid Shapley-Shubik power indices for any n is beneficial to solving the problem for $n = 4$. The following theorem only uses properties of the addition of nonnegative integers and permutations.

Theorem 1: *If α_k is the pivotal voter of the permutation $\alpha_1 \alpha_2 \ldots \alpha_n$ of voters, then voter α_k's power index is at least $(k-1)!(n-k)!$*

Proof: Let $w(\alpha_k)$ be the weight of voter α_k. Assume that k satisfies $1 < k < n$. The permutation can be written as $\alpha_1 \alpha_2 \ldots \alpha_{k-1} \alpha_k \alpha_{k+1} \ldots \alpha_n$. Because α_k is the pivotal voter, $\sum_{i=1}^{k-1} w(\alpha_i) < q$ and $\sum_{i=1}^{k} w(\alpha_i) \geq q$. It follows that α_k is the pivotal voter for the permutation of voters $\beta_1 \beta_2 \ldots \beta_{k-1} \alpha_k \gamma_{k+1} \ldots \gamma_n$, where $\beta = \beta_1 \ldots \beta_{k-1}$ is a permutation of $\{\alpha_1, \ldots, \alpha_{k-1}\}$ and $\gamma = \gamma_{k+1} \ldots \gamma_n$ is a permutation of $\{\alpha_{k+1}, \ldots, \alpha_n\}$. There are $(k-1)!$ permutations β and $(n-k)!$ permutations γ. Hence, α_k is the pivotal voter in at least $(k-1)!(n-k)!$ permutations.

If $k = 1$, then α_1 is the pivotal voter in $\alpha_1 \alpha_2 \ldots \alpha_n$. It follows that $w(\alpha_1) \geq q$ and that α_1 is a dictator. This implies that α_1 is the pivotal voter for all $n!$ permutations of the n voters. And, for $k = 1$, $n! > (k-1)!(n-k)!$ is in agreement with the theorem.

For $k = n$, the voter α_n is pivotal only when he appears at the end of the sequence of voters, or equivalently, at the end of the permutation. This occurs $(n-1)!$ times which is equal to $(k-1)!(n-k)!$ for $k = n$.

This theorem can be applied to eliminate certain potential image points of the Shapley-Shubik power indices.

3.4 Integrating Combinatorics, Geometry, and Probability through the Shapley-Shubik Power Index

Corollary 1: *For n voters, a non-dummy voter's Shapley-Shubik power index is at least $(\lceil \frac{n}{2} \rceil - 1)!(n - \lceil \frac{n}{2} \rceil)!$ where $\lceil \cdot \rceil$ represents the ceiling function.*

Corollary 2: *For $n \geq 4$ voters, no voter has an odd power index.*

Proof: By Theorem 1, a non-dummy, non-dictator voter's power index is the sum of terms $(k-1)!(n-k)!$. For $n \geq 4$, $(k-1)!(n-k)!$ is even. And, the sum of even numbers is even. And, dictators and dummies always have even power indices.

Exercise 18: Describe which possible Shapley-Shubik image points are eliminated by the corollaries for $n = 4$?

3.4.4 Discrete Approach to Probabilistic Questions

Re-examine Figure 3.4.1, where the Shapley-Shubik power index is determined for $[3; 2, 1, 1]$. The inequalities present in the figure determine which voter is pivotal for a particular sequence of voters. The set of all inequalities defines the Shapley-Shubik power index. Indeed, any simple weighted-voting game $[q; w_1, w_2, w_3]$ with $w_1 + w_2 + w_3 = w$ satisfying

$$w_1 < q \quad w_2 < q \quad w_3 < q$$

$$w_1 + w_2 \geq q \quad w_1 + w_3 \geq q \quad w_2 + w_3 < q$$

$$w_1 + w_2 + w_3 \geq q$$

will have the *same* Shapley-Shubik power index, $4 : 1 : 1$, as $[3; 2, 1, 1]$.

Definition 4: For fixed w and q, two n-voter simple weighted-voting games are *ss*-equivalent if they have the same Shapley-Shubik power index.

Exercise 19: Show that the set of all *ss*-equivalent, n-voter simple weighted-voting games form an equivalence class.

It becomes a counting problem to determine how many simple weighted-voting games with fixed w and q are in any equivalence class, as demonstrated by the following two examples.

Example 3: Determine the number of simple weighted-voting games of the form $[14; w_1, w_2, w_3]$ with $w = 20$ in the equivalence class of $6 : 0 : 0$.

Since voter 1 is a dictator, it follows that $w_1 \geq 14$. The number of simple weighted-voting games $[14; w_1, w_2, w_3]$ with $w = 20$ in the equivalence class of $6 : 0 : 0$ is the number of nonnegative integer solutions of $w_1 + w_2 + w_3 = 20$ where $w_1 \geq 14$. Equivalently, it is $\sum_{w_1=14}^{20} |S_{w_1}|$ where S_{w_1} is the set of nonnegative integer solutions to $w_2 + w_3 = 20 - w_1$. Using a simple counting argument, it follows that $|S_{w_1}| = \binom{20-w_1+1}{1} = 20 - w_1 + 1 = 21 - w_1$. Hence, the number of simple weighted-voting games $[14; w_1, w_2, w_3]$ with $w = 20$ in the equivalence class of $6 : 0 : 0$ is

$$\sum_{w_1=14}^{20} |S_{w_1}| = \sum_{w_1=14}^{20} 21 - w_1 = 7 \cdot 21 - \sum_{w_1=1}^{20} w_1 + \sum_{w_1=1}^{13} w_1 = 147 - 210 + 91 = 28.$$

The next example demonstrates the counting problem for the set of inequalities from Figure 3.4.1.

Example 4: Determine the number of simple weighted-voting games of the form $[14; w_1, w_2, w_3]$ with $w = 20$ in the equivalence class of $4 : 1 : 1$.

Determining the number of simple weighted-voting games in the equivalence class is equivalent to determining the number of nonnegative integer solutions that satisfy $w_1 + w_2 + w_3 = 20$ and the following inequalities

$$w_1 < 14 \quad w_2 < 14 \quad w_3 < 14$$

$$w_1 + w_2 \geq 14 \quad w_1 + w_3 \geq 14 \quad w_2 + w_3 < 14.$$

Since $w_1 + w_2 + w_3 = 20$, the inequalities $w_1 + w_2 \geq 14$, $w_1 + w_3 \geq 14$, and $w_2 + w_3 < 14$ can be rewritten as $6 \geq w_3$, $6 \geq w_2$, and $6 < w_1$, respectively. But, $6 \geq w_3$ and $6 \geq w_2$ imply that $w_1 \geq 8$. Hence, we are counting the number of nonnegative integer solutions of $w_1 + w_2 + w_3 = 20$ such that $8 \leq w_1 \leq 13$, $w_2 \leq 6$, and $w_3 \leq 6$.

Let $w_1^* = w_1 - 8$, $w_2^* = w_2$, and $w_3^* = w_3$. The system transforms to $w_1^* + w_2^* + w_3^* = 12$ such that $0 \leq w_1^* \leq 5$, $0 \leq w_2^* \leq 6$, and $0 \leq w_3^* \leq 6$. Determining the number of integer solutions to the equality satisfying the inequality constraints can be achieved through a use of the Inclusion-Exclusion Principle (*e.g.,* see Brualdi [3]). Let S be the set of nonnegative integer solutions of $w_1^* + w_2^* + w_3^* = 12$. Further, let P_1 be the property that $w_1^* > 5$ and let P_i be the property that $w_i^* > 6$, for $i = 2$ and 3. Define the set $A_i = \{(w_1^*, w_2^*, w_3^*) : (w_1^*, w_2^*, w_3^*) \in S$ and (w_1^*, w_2^*, w_3^*) has property $P_i\}$. Then, the number of nonnegative integer solutions of $w_1^* + w_2^* + w_3^* = 12$ such that $w_1^* \leq 5$, $w_2^* \leq 6$, and $w_3^* \leq 6$ is, by the Inclusion-Exclusion Principle,

$$|A_1' \cap A_2' \cap A_3'| = |S| - |A_1| - |A_2| - |A_3| + |A_1 \cap A_2| + |A_1 \cap A_3| + |A_2 \cap A_3| - |A_1 \cap A_2 \cap A_3|$$

where A' is the complement of set A. This simplifies to $\binom{14}{2} - \binom{8}{2} - \binom{7}{2} - \binom{7}{2} = 21$. There are 21 simple weighted-voting games with $w = 20$ and $q = 14$ *ss*-equivalent to $4 : 1 : 1$.

Assume that there is a uniform distribution over all simple weighted-voting games with w and q fixed. That is, all simple weighted-voting games

$[q; w_1, w_2, \ldots, w_n]$ such that $w_1 + w_2 + \cdots + w_n = w$ are equally likely to occur.

Then the size of the equivalence class of the simple weighted-voting games for a Shapley-Shubik power index naturally corresponds to the likelihood of picking a simple weighted-voting game at random which has that particular Shapley-Shubik power index. Consider the following extension of the previous example.

Example 5: Determine the likelihood of a simple weighted-voting game having Shapley-Shubik power index $4 : 1 : 1$ for $w = 20$ and $q = 14$.

There are $\binom{22}{2} = 231$ simple weighted-voting games $[14; w_1, w_2, w_3]$ with $w_1 + w_2 + w_3 = w = 20$ that have Shapley-Shubik power index of $4 : 1 : 1$; this is the same number of nonnegative integer solutions of $w_1 + w_2 + w_3 = w = 20$. If all of these games are equally likely to occur, then the probability of selecting any one at random is $\frac{1}{231}$. Hence, the probability of a simple weighted-voting game with $w = 20$ and $q = 14$ having Shapley-Shubik power index is $\frac{21}{231} = \frac{1}{11}$, since there are 21 simple weighted-voting games with $w = 20$ and $q = 14$ with Shapley-Shubik power index $4 : 1 : 1$.

The idea of considering the probabilities in which certain Shapley-Shubik power indices occur is a natural extension of the counting problem. Consider the possibility that there are a fixed number of shares of stock of a company, but that shareholders may sell their stock to other shareholders. We assume that all arrangements are equally possible. There are examples and exercises in COMAP [4] that consider how many shares could be sold without changing the power index of the game.

Exercise 20: Determine how many 3-voter, simple weighted-voting games with $w = 20$ and $q = 14$ are *ss*-equivalent to $3 : 3 : 0$.

3.4 Integrating Combinatorics, Geometry, and Probability through the Shapley-Shubik Power Index   151

Exercise 21: Find a representative simple weighted-voting game with 4 voters that has Shapley-Shubik power index $6:6:6:6$ for $w = 20$ and $q = 11$. Compare this to a representative 4-voter game that has Shapley-Shubik power index $6:6:6:6$ for $w = 20$ and $q = 18$. Discuss the relationships between the pivotal voters in the two games.

3.4.5 Geometrical Interpretation of the Shapley-Shubik Power

A simple weighted-voting game $[q; w_1, w_2, \ldots, w_n]$ can be normalized by dividing the quota and players' weights by $w = w_1 + w_2 + \cdots + w_n$. The players' weights of the normalized game

$$\left[\frac{q}{w}; \frac{w_1}{w}, \frac{w_2}{w}, \ldots, \frac{w_n}{w}\right]$$

can be considered geometrically as a point on the $(n-1)$-simplex, denoted S_{n-1}; the $(n-1)$-simplex is the set of nonnegative solutions to

$$x_1 + x_2 + \cdots + x_n = 1.$$

Since any $n - 1$ values of x_i's define a point on the simplex, the simplex has dimension $n - 1$. For a game with 3 players, the normalized weights of the three players can be viewed as a point on the 2-simplex. The 2-simplex is the intersection of the plane $x_1 + x_2 + x_3 = 1$ and the positive octant where $x_i \geq 0$ for all i; this forms a triangle as shown in Figure 3.4.2.

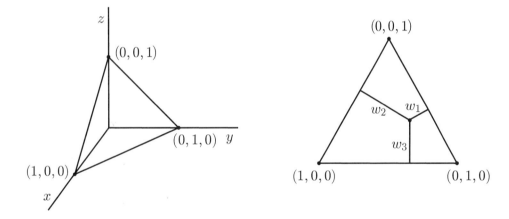

Figure 3.4.2.

Up to now, we have restricted our attention to discrete simple weighted-voting games where the weights were nonnegative integer values. In this section, we allow weights to be nonnegative real numbers; we call these *continuous simple weighted-voting games*. For the quota being a fixed percentage of the sum weight of all voters, a point on S_{n-1} represents an equivalence class of simple weighted-voting games for n players. This follows because many simple weighted-voting games have the same normalized representation. The simple weighted-voting game $[q; w_1, w_2, \ldots, w_n]$ is normalized to $\left[\frac{q}{w}; \frac{w_1}{w}, \frac{w_2}{w}, \ldots, \frac{w_n}{w}\right]$. The weights of the pre-normalized game can be viewed as the point $(w_1, w_2, \ldots, w_n)$ in R^n. The normalized weights are represented by the point $\left(\frac{w_1}{w}, \frac{w_2}{w}, \ldots, \frac{w_n}{w}\right)$ on the $(n-1)$-simplex. Geometrically, these two points are related since the line connecting the origin and $(w_1, w_2, \ldots, w_n)$ intersects the $(n-1)$-simplex at $\left(\frac{w_1}{w}, \frac{w_2}{w}, \ldots, \frac{w_n}{w}\right)$. Hence, the equivalence class of simple weighted-voting games represented by $\left(\frac{w_1}{w}, \frac{w_2}{w}, \ldots, \frac{w_n}{w}\right)$ is every point on the ray from the origin through $\left(\frac{w_1}{w}, \frac{w_2}{w}, \ldots, \frac{w_n}{w}\right)$.

The Shapley-Shubik power index of a simple weighted-voting game of n voters, $s_1 : s_2 : \cdots : s_n$, also can be normalized. Specifically, the normalized power index $\frac{s_1}{n!} : \frac{s_2}{n!} : \cdots : \frac{s_n}{n!}$ is a point on the

$(n-1)$-simplex since

$$\frac{s_1}{n!} + \frac{s_2}{n!} + \cdots + \frac{s_n}{n!} = 1$$

where $\frac{s_j}{n!} \geq 0$ for all j. The 10 possible Shapley-Shubik power indices for 3-voter simple weighted-voting games are 6:0:0, 0:6:0, 0:0:6, 4:1:1, 1:4:1, 1:1:4, 3:3:0, 3:0:3, 0:3:3, and 2:2:2 (this answers Exercise 8). Normalized, these power indices are graphed on the 2-simplex in Figure 3.4.3; they are $(1, 0, 0)$, $(0, 1, 0)$, $(0, 0, 1)$, $(\frac{2}{3}, \frac{1}{6}, \frac{1}{6})$, $(\frac{1}{6}, \frac{2}{3}, \frac{1}{6})$, $(\frac{1}{6}, \frac{1}{6}, \frac{2}{3})$, $(\frac{1}{2}, \frac{1}{2}, 0)$, $(\frac{1}{2}, 0, \frac{1}{2})$, $(0, \frac{1}{2}, \frac{1}{2})$, and $(\frac{1}{3}, \frac{1}{3}, \frac{1}{3})$.

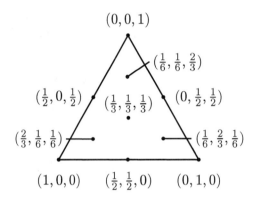

Figure 3.4.3.

For the remainder of the paper we will assume that all weights, quotas, and Shapley-Shubik power indices are normalized, unless otherwise indicated. Then the normalized quota is in $\left(\frac{1}{2}, 1\right]$. The Shapley-Shubik power index can be viewed as a map from S_{n-1} to S_{n-1}, mapping a simple weighted-voting game (with a specified normalized quota q) to a Shapley-Shubik power index. We will denote this map by P_q. As discussed previously (*e.g.*, in Figure 3.4.1), the Shapley-Shubik power index of a given game $[q; w_1, w_2, \ldots, w_n]$ (assumed to be normalized) is determined by which sums of weights are greater than or equal to the quota. The collection of inequalities partition S_{n-1} into different regions where each point in a region has the same Shapley-Shubik power index.

Each point in a specific partition has the same Shapley-Shubik power index; thus, every point in a region has the same image under P_q. Graphing the boundaries of the regions, *i.e.*, all combinations of sums of weights equaling the quota, yields a geometric representation of the partitioning of the simplex by the Shapley-Shubik power index map.

For $n = 3$, the partitioning equations of S_2 are

$$w_1 = q, \quad w_2 = q, \quad w_3 = q, \quad w_1 = 1 - q, \quad w_2 = 1 - q, \text{ and } w_3 = 1 - q.$$

Due to the symmetry, there are 4 regions up to permutation on the set of players; these regions are described below. Summative data appears in Table 3.4.1 and the regions are pictured in Figure 3.4.4. Note that the shape of the regions is dependent on the quota. For the below descriptions, assume that the normalized quota is fixed.

Exercise 22: From Figure 3.4.1, it appears that $w_1 + w_2 = q$ should be one of the equations that partition the 2-simplex. Explain why $w_1 + w_2 = q$ is accounted for in the previous paragraph.

Case 1. (Dictator Regions R_1, R_2, and R_3). A point (w_1, w_2, w_3) is in region R_i if and only if voter i is a dictator, i.e., $w_i \geq q$. All of the points in R_i are mapped to the Shapley-Shubik power index with a 1 in position i and 0 elsewhere.

Case 2. (Regions R_4, R_5, and R_6). A point (w_1, w_2, w_3) is in region R_{i+3} if and only if voter i is a dummy voter and the other two voters have equal power. This occurs when $w_i + w_j < q$, $w_i + w_k < q$,

3.4 Integrating Combinatorics, Geometry, and Probability through the Shapley-Shubik Power Index   153

and $w_j + w_k \geq q$ where i, j, and k are distinct voters in $\{1, 2, 3\}$. All of the points in region R_{i+3} are mapped to the Shapley-Shubik power index with a 0 in position i and $\frac{1}{2}$ in the other two positions.

Case 3. (Regions R_7, R_8, and R_9). A point (w_1, w_2, w_3) is in region R_{i+6} if and only if $w_i < q$, $w_i + w_j \geq q$, $w_i + w_k \geq q$, and $w_j + w_k < q$. All of the points in region R_{i+6} are mapped to the Shapley-Shubik power index with a $\frac{2}{3}$ in position i and $\frac{1}{6}$ in the other two positions.

Case 4. (Region R_{10}). A point (w_1, w_2, w_3) is in region R_{10} if all voters have equal power. This can occur in different ways depending on the quota. For $q \leq \frac{2}{3}$, a point is in this region if $w_i \leq 1 - q$ for all voters i. Equivalently, $w_j + w_k \geq q$ for all voters j and k. For $q > \frac{2}{3}$, a point is in this region if $w_i + w_j < q$ for all voters i and j. Hence, a measure passes only if all three voters agree in the affirmative. All of the points in region R_{10} are mapped to the Shapley-Shubik power index $(\frac{1}{3}, \frac{1}{3}, \frac{1}{3})$.

The above lengthy reasoning amounts to saying that any game lying in region R_i has the Shapley-Shubik power index as listed in Table 3.4.1.

Region	R_1	R_2	R_3	R_4	R_5	R_6	R_7	R_8	R_9	R_{10}
SSPI	$1:0:0$	$0:1:0$	$0:0:1$	$0:\frac{1}{2}:\frac{1}{2}$	$\frac{1}{2}:0:\frac{1}{2}$	$\frac{1}{2}:\frac{1}{2}:0$	$\frac{2}{3}:\frac{1}{6}:\frac{1}{6}$	$\frac{1}{6}:\frac{2}{3}:\frac{1}{6}$	$\frac{1}{6}:\frac{1}{6}:\frac{2}{3}$	$\frac{1}{3}:\frac{1}{3}:\frac{1}{3}$

Table 3.4.1. Regions and their corresponding Shapley-Shubik power indices.

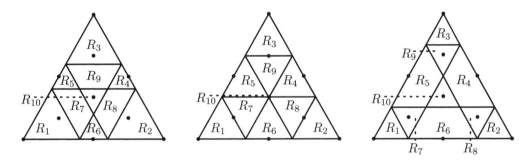

Figure 3.4.4.

Exercise 23: Normalize the simple weighted-voting game $[21; 7, 12, 13]$. Which region contains the normalized game? Determine the Shapley-Shubik power index of the game.

Exercise 24: Normalize the simple weighted-voting game $[3; 2, 1, 1]$. Which region contains the normalized game? Determine the Shapley-Shubik power index of the game.

Exercise 25: For $q = \frac{3}{4}$, find the Shapley-Shubik power index of the game corresponding to the point given in Figure 3.4.5.

Example 6: The normalized game from Example 3 is $[\frac{14}{20}; \frac{w_1}{20}, \frac{w_2}{20}, \frac{w_3}{20}]$. As in Example 3, consider the possible values of w_1, w_2, and w_3 so that the simple weighted-voting game has a Shapley-Shubik power index of $\frac{2}{3}:\frac{1}{6}:\frac{1}{6}$. There are a finite number of games with $w = 20$ (before normalization). Normalized, these games form a set of lattice points on the 2-simplex. The games that have the same Shapley-Shubik power index will lie in the same partition of the simplex.

The Shapley-Shubik power index of $\frac{2}{3}:\frac{1}{6}:\frac{1}{6}$ corresponds to region R_7 in Figure 3.4.4. Thus, Example 6 can be viewed as counting the lattice points that lie within region R_7. All 231 integer solutions to $\frac{w_1}{20} + \frac{w_2}{20} + \frac{w_3}{20} = 1$ are evenly spaced out in the simplex and are pictured in Figure 3.4.6. The 21 lattice

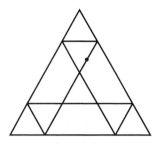

Figure 3.4.5.

points in region R_7 are darkened in Figure 3.4.6; this agrees with the 21 lattice points from the solution to Example 6.

In Example 6, if w is increased, then the total number of discrete games increases since the number of solutions to $w_1 + w_2 + w_3 = w$ increases. Indeed, there are $\binom{w+2}{2} = \frac{(w+2)(w+1)}{2}$ nonnegative integer solutions to $w_1 + w_2 + w_3 = w$. Consequently, the number of lattice points on the 2-simplex increases. As w increases without bound, the lattice points fill up the simplex. In particular, the lattice points that have the Shapley-Shubik power index of $\frac{2}{3} : \frac{1}{6} : \frac{1}{6}$ fill up region R_7. Although the number of games in region R_7 increases without bound as $w \to \infty$, the ratio of games in region R_7 to the total number of games approaches the ratio of the area of region R_7 to the area of the 2-simplex.

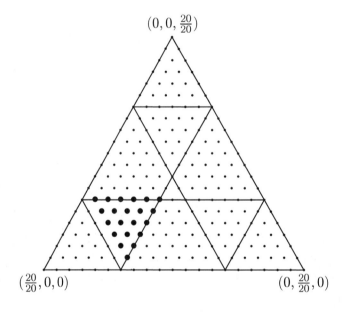

Figure 3.4.6.

Assume that there exists a uniform distribution over the n-simplex. So every point is equally likely to be selected as the weights of a simple weighted-voting game. For a fixed quota q, the likelihood that a particular Shapley-Shubik power index occurs is merely the volume of the region that it maps to divided by the volume of the simplex. Although it is difficult to picture the higher dimensional simplices, even for $n = 4$, it is still possible to determine the likelihood of certain outcomes. For $n = 3$, it is quite easy to compute the probability of a simple weighted-voting game having a particular Shapley-Shubik power index. These probabilities are computed as functions of the quota in Table 3.4.2. We consider this limit process explicitly in the next example.

3.4 Integrating Combinatorics, Geometry, and Probability through the Shapley-Shubik Power Index

Region	R_1, R_2, R_3	R_4, R_5, R_6	R_7, R_8, R_9	R_{10}
Probability ($q < \frac{2}{3}$)	$(1-q)^2$	$(2q-1)^2$	$-8q^2 + 10q - 3$	$(2-3q)^2$
Probability ($q = \frac{2}{3}$)	$\frac{1}{9}$	$\frac{1}{9}$	$\frac{1}{9}$	0
Probability ($q > \frac{2}{3}$)	$(1-q)^2$	$-5q^2 + 8q - 3$	$(1-q)^2$	$(3q-2)^2$

Table 3.4.2. Probabilities of regions for all values of the quota.

Exercise 26: Assume the normalized quota is $q < \frac{2}{3}$. Use basic geometry to verify the probability that a simple weighted-voting game selected at random from a uniform distribution over the 2-simplex has $0 : \frac{1}{2} : \frac{1}{2}$ as its Shapley-Shubik power index is $(2q-1)^2$. (*Hint*: Remember to divide the area of R_4 by the area of the 2-simplex.)

Exercise 27: For each of $q < \frac{2}{3}$, $q = \frac{2}{3}$, and $q > \frac{2}{3}$, verify that the sum of the probabilities that each region occurs is 1.

Example 7: Assume that the normalized quota q is greater than $\frac{2}{3}$ and that w is an integer. As w approaches infinity, we can determine the likelihood that a simple weighted-voting game $[q; \frac{w_1}{w}, \frac{w_2}{w}, \frac{w_3}{w}]$ selected at random has Shapley-Shubik power index $\frac{2}{3} : \frac{1}{6} : \frac{1}{6}$. As w increases, we expect that this likelihood will converge to $(1-q)^2$, the value from Table 3.4.2 for region R_7.

The inequalities that define region R_7 (from Case 3) are

$$\frac{w_1}{w} < q, \quad \frac{w_1 + w_2}{w} \geq q, \quad \text{and} \quad \frac{w_1 + w_3}{w} \geq q.$$

Using a little algebra, these inequalities can be rewritten as

$$w_1 < qw, \quad w_3 \leq (1-q)w, \quad \text{and} \quad w_2 \leq (1-q)w.$$

Because $w_3 \leq (1-q)w$ and $w_2 \leq (1-q)w$, it follows that $w_2 + w_3 \leq 2(1-q)w$ or $w_1 + w_2 + w_3 \leq 2(1-q)w + w_1$. This simplifies to $(2q-1)w \leq w_1$. This bounds w_1 from below. The number of simple weighted-voting games in region R_7 for fixed w is the number of nonnegative integer solutions of

$$w_1 + w_2 + w_3 = w \text{ subject to}$$
$$(2q-1)w \leq w_1 < qw$$
$$w_2 \leq (1-q)w$$
$$w_3 \leq (1-q)w.$$

For different values of w, the products qw and $(1-q)w$ may or may not be integers. Since we are concerned with the limit process, we will assume that qw and $(1-q)w$ are always integer values. This assumption does not change the calculation. Indeed, one could look at the case where these values are integers or are not integers. In both cases, the limits converge to the same value.

As in Example 1, we can use the Inclusion Exclusion Principle to count the number of nonnegative integer solutions to the above equality with inequality constraints. However, it is simpler to transform the system above by a change of variables, as would be done in the Inclusion-Exclusion Principle, but then consider the transformed system geometrically. Let $w_1^* = w_1 - (2q-1)w$, $w_2^* = w_2$, and $w_3^* = w_3$. The above system becomes

$$w_1^* + w_2^* + w_3^* = 2(1-q)w \quad \text{s.t.} \quad w_1^* < (1-q)w \quad \text{and} \quad w_i^* \leq (1-q)w \quad \text{for } i = 2, 3.$$

Hence, the number of nonnegative integer solutions to this new system is equal to the number of solutions to the original system.

Notice that all of the inequalities of the new system involve $(1-q)w$, exactly half of $2(1-q)w$, the sum of the variables. This allows us to easily visualize the solutions, as in Figure 3.4.7. The darkened dots represent solutions; this region forms an equilateral triangle of lattice points with $(1-q)w$ points on every side. So, the number of solutions is the sum of the first $(1-q)w$ positive integers. Hence, there are $\binom{(1-q)w+1}{2}$ solutions.

To determine the ratio of solutions to possible simple-weighted voting games, we merely divide $\binom{(1-q)w+1}{2}$ by the number of simple-weighted voting games with total weight w. There are $\binom{w+2}{2}$ such games. Hence, the limit of the number of games that have Shapley-Shubik power index of $\frac{2}{3} : \frac{1}{6} : \frac{1}{6}$ is

$$\lim_{w \to \infty} \frac{[(1-q)w][(1-q)w+1]}{(w+2)(w+1)} = \lim_{w \to \infty} \frac{(1-q)^2 w^2 + (1-q)w}{w^2 + 3w + 2} = (1-q)^2.$$

This value agrees with the entry in Table 3.4.2 for region R_7 when $q > \frac{2}{3}$.

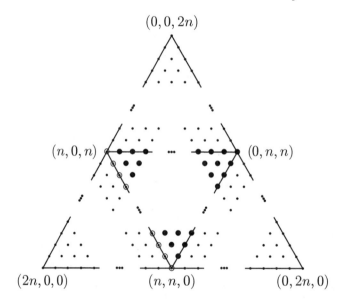

Figure 3.4.7.

Exercise 28: Use Table 3.4.2 to determine the probability that a randomly selected, continuous simple weighted-voting game with normalized quota $q = 0.7$ will have a Shapley-Shubik power index of $0 : \frac{1}{2} : \frac{1}{2}$.

Exercise 29: For normalized quota $q = 0.7$, determine the number of discrete simple weighted-voting games with a Shapley-Shubik power index of $0 : \frac{1}{2} : \frac{1}{2}$ and $w = 20$. Now determine the probability that a randomly chosen discrete simple weighted-voting game with normalized quota 0.7 and $w = 20$ has a Shapley-Shubik power index of $0 : \frac{1}{2} : \frac{1}{2}$. If w were increased to 100, how would you expect the probability to change?

We have linked the discrete and continuous simple weighted-voting games through limits. We have also come full circle. Our initial mathematics questions were motivated by the modeling of political interactions.

We now have new tools to ask and answer questions about the likelihood of different outcomes. As an example, the next proposition applies the information from Table 3.4.2 and considers it in context of the modeling of political science. The corollary is a direct consequence of the proof of the proposition.

Proposition 1: *The probability of voter i being a dictator in a 3-voter simple weighted-voting game with normalized quota q is $(1-q)^2$.*

Proof: Let $i = 1$. Voter 1 is a dictator if and only if $w_1 \geq q$. The region on the simplex satisfying $w_1 \geq q$ is a triangle similar to the 2-simplex where the two triangles share the vertex $(1, 0, 0)$. The intersection of the hyperplane $w_1 = q$ and the simplex is the line segment with endpoints $(q, 1-q, 0)$ and $(q, 0, 1-q)$. The ratio of corresponding sides of the smaller simplex to the larger is $\sqrt{2}(1-q) : \sqrt{2}$ or $(1-q)$. By Euclid VI.19, the ratio of the areas of the smaller to the larger simplices is $(1-q)^2$. By symmetry, the proposition is true for $i = 1, 2,$ or 3.

Corollary: *The probability of there being a dictator in a 3-voter simple weighted-voting game with normalized quota q is $3(1-q)^2$.*

The proposition can be extended for any n using the relationship between the volume of similar n-dimensional regions. Haines and Jones [8] contains the extension and applications of the techniques and perspectives of this paper to power indices and apportionment methods.

3.4.6 Guidelines for Use

This article naturally begins at an introductory level and offers more challenging aspects as it evolves. As a rule of thumb, earlier material has been used in multiple settings: a general education requirement course, an undergraduate applied combinatorics and graph theory course, and a graduate course in combinatorics, as well as in independent study courses. As the material becomes more difficult, it is accessible to fewer students. The following descriptions offer our suggestions on how to use this material for different courses, including the aforementioned, as well as in a course in mathematical modeling. These guidelines are based on adaptations of this material in different level courses at Montclair State University from Spring 1999 to Spring 2002.

General Education Requirement Course

For a general education requirement course or mathematics for liberal arts course, the topics and treatment introducing simple weighted-voting games and Shapley-Shubik power indices, as well as which power indices are possible, are appropriate. We suggest beginning with specific stories about political processes and asking how such processes might be modeled and who has the political power. After defining simple weighted-voting games, the students should be able to handle the subtleties of the constraint condition on the quota, as considered in Exercise 2. In such courses, the primary goal should be for the students to understand the range of applications of mathematics. For this reason, it is important to focus on modeling real and fictitious scenarios. Indeed, students should be asked to model different political institutions, such as the US Congress, the European Union, etc. Students can easily understand aspects of modeling, such as how changes in weights must be accompanied by a change in the quota, as in Exercise 3.

Students can become quite adept at translating the words and descriptions of a political process (specifically, what constitutes a coalition that can pass a measure) into mathematical conditions and ultimately simple weighted-voting games. One such problem is Exercise 4. Students should realize that there is more than one way to model the same situation. This is a good opportunity to discuss the idea of two simple weighted-voting games being equivalent, having the same sets of players being winning, losing, and blocking coalitions. Students also can provide the conditions for a measure to pass if they are given a simple weighted-voting game, although we do not provide any exercises of this form. Such examples can be found in Lampert [12] and COMAP [4]. The definitions of a dictator, veto power, and a dummy voter are easy for students to understand. Translating these concepts into mathematics after defining the Shapley-Shubik power index is a little more difficult for students, but definitely doable. For example, Exercises 8 and 9 can be approached by first considering the case where there are 3 voters. Students can usually get the results for $n = 3$ alone or in a group, but may need some assistance to make the jump for general n. More generally, students should be able to understand that a player whose vote has a larger weight will be at least as powerful (as defined by the Shapley-Shubik Power Index) as a player whose vote has a smaller weight. Of course, when introducing the Shapley-Shubik power index, there are nice opportunities to discuss and motivate permutations and factorial notation.

Although the more rigorous definition of the Shapley-Shubik power index as a function is inappropriate for most general education courses, some of the exercises and concepts in the third section are appropriate if pitched at the right level. By having the students focus on the patterns of circles that can occur when determining the pivotal voter for all permutations of voters (as demonstrated in Figure 3.4.1), students can readily answer Exercises 12 and 13. After tallying the results from different examples where there are 3 voters, it is reasonable to focus on the possible indices that do not appear to represent the power of players in a simple weighted-voting game. The required analysis is a little more subtle, but can be accomplished by students in a mathematics for liberal arts course. For example, Exercise 15 should not be too difficult. It may be useful to explain why 3:2:1 is not a valid Shapley-Shubik power index (Example 1).

Overview for Advanced Courses Including Modeling

For more advanced courses, the material that is covered in a general education course is still applicable; it can be covered much more quickly and efficiently. If the advanced course is a course in mathematical modeling, more time should be spent on modeling different situations. The guidelines for implementing this material in an upper level applied combinatorics course are suitable for those interested in adapting this material to a class in modeling. However, there are more opportunities to challenge the students by having them read material from works in mathematical political science that model different political institutions and then discuss how the mathematics indicates who has power in a political institution. The end of the introduction cites appropriate material for modeling students to read, with some assistance.

Undergraduate Applied Combinatorics

For an undergraduate applied combinatorics course, simple weighted-voting games and the Shapley-Shubik power index provide a thread that can weave through the course as new topics are introduced. In fact, the desire to consider different aspects of simple weighted-voting games and power indices can be used to motivate different topics. There is some flexibility about the order of presentation of the topics. The presentation in this article introduces permutations, equivalence relations, and the Inclusion-Exclusion Principle to determine the number of nonnegative integer solutions to equalities with inequality constraints. The geometrical perspectives of viewing nonnegative integer solutions of equalities as points on a simplex and connecting probabilities to the area of regions in the simplex are not standard topics in a combinatorics course, but they do add insight to the analysis of the equivalence classes and the Inclusion-Exclusion Principle. See Table 3.4.3 for a list of combinatorics content and the relevant material in simple weighted-voting games and the Shapley-Shubik power index. We offer time estimates that assume that other topics will also be covered in the class. Our estimates describe how many days the content was taught as part of the lesson; we assume that class periods are 75 minutes in duration.

An appropriate place to introduce simple weighted-voting games and the Shapley-Shubik power index is when combinations and permutations are covered, hopefully early in the course. Combinations are naturally considered when evaluating winning coalitions. At this point, the structure of voters necessary to pass a measure can be used to discuss equivalence classes informally, since different values for quota and weights can yield the same structure. Although not considered in this article, at this point it is possible and natural to introduce the Banzhaf power index that is based on combinations, as opposed to permutations, of voters (*e.g.*, see [4]). By keeping track of the Shapley-Shubik power indices that are associated with different games, one can lead students to question which power indices are possible. Then, using properties of the addition of real numbers and permutations, it is possible to prove that 5:1:0 is not a valid Shapley-Shubik power index and ask the class to prove or disprove that 3:2:1 is a valid Shapley-Shubik power index. Parts of two class periods can be used to introduce simple weighted-voting games and the Shapley-Shubik power index. More time should be spent if the focus of the course is modeling.

Students always seem to have difficulties defining functions and relations, especially in discrete settings. For example, students often exclaim "vertical line test" when trying to determine whether a relation is a function in a discrete setting. Viewing the Shapley-Shubik power index as a function with appropriately defined domain and range demonstrates how mathematical structure can help formalize questions that arise in the application. From examples, students will realize that the Shapley-Shubik power index function is, for the appropriate parameter values, many-to-one. This observation can be translated into questions about the number of points in the domain that have the same image. Of course, this leads back to equivalence classes in the language of functions.

Although the Shapley-Shubik power index can be used to introduce the idea of an equivalence relation, there is no reason to cover this material immediately. Before introducing partial orderings and equivalence relations, it is useful to teach the use of algorithms to generate combinations and permutations, as well as inversion sequences. It also helps to define simple weighted-voting games to be *ss*-equivalent, as is done in this article. Then, the students can help the instructor prove that *ss*-equivalent games form an equivalence class. It may be worth considering all possible distributions of weights of voters that sum to a fixed sum and yield the same Shapley-Shubik power index. This will lead to partitioning the nonnegative integer solutions to the simplex equation with the inequality constraints. Alternatively, it can be equally productive to explicitly demonstrate the link between computing the power index and the equation with the inequality constraints. Changing the order of presentation would then lead into the *ss*-equivalent definition.

Generalizing the equality with inequality constraints for more than three voters can be discussed, covered in detail, or even omitted, depending on the time set aside for this application and the familiarity or comfort of the professor with the topics. However, it is natural to consider what consequences follow for simple weighted-voting games with more than three voters. The extension to more than three voters does raise interesting complexity-type questions about the number of inequality constraints necessary to form an equivalence class. While the definitions, theorem, etc. hold for simple weighted-voting games with any number of players, things like computing examples and determining what points in the domain of the Shapley-Shubik power index are actual images can be tedious.

There are many variations on the theme of determining the number of simple weighted-voting games that are in the same equivalence class (and have the same Shapley-Shubik power index). However, in all such cases, it is necessary to emphasize setting up the system of inequality constraints and to indicate how manipulating the constraints assists in finding the minimal constraints. As is demonstrated in this article, some of the constraints may end up being redundant (as in Example 4). Once the inequality constraints are determined, the problem becomes solvable by counting methods. Example 4 demonstrates how changes of variable simplify the calculations. Many of these problems will require the use of the Inclusion-Exclusion Principle. We suggest that the Inclusion-Exclusion Principle as well as determining the number of nonnegative integer solutions to an equality with inequality constraints be introduced before

Pertinent Topic	Relevant Voting Theory Content
Combinations	Winning and losing coalitions for simple weighted-voting games Number of coalitions of fixed size Banzhaf power index
Permutations	Shapley-Shubik power index Number of possible power indices Introduction to simplices
Discrete functions	Shapley-Shubik power index as a function Many-to-one, injective, and surjective functions Informal introduction to partitions
Equivalence Classes	Formal partitions Definition of ss-equivalent games Geometric partition of the simplex
Nonnegative integer solutions	Equations with inequality constraint view of ss-equivalence Counting the number of solutions Geometry of the solutions
Inclusion-Exclusion Principle	Counting the number of nonnegative integer solutions Solutions to equalities with inequality constraints
Large n	Areas of regions defining a partition of the simplex Probabilities of different Shapley-Shubik power index outcomes

Table 3.4.3. Topics in an undergraduate combinatorics course and their relationship to simple weighted-voting games and the Shapley-Shubik power index.

determining the number of simple weighted-voting games in an ss-equivalence class. One class period can be used to introduce and to apply the Inclusion-Exclusion Principle to count the number of discrete games in an equivalence class.

Although the counting problems can be taught without introducing the geometric perspective, it is most helpful to make the geometry the focus of instruction. The inequality constraints that determine if a simple weighted-voting game is in a particular equivalence class have obvious geometrical implications even though it may not be all that obvious when mixing in the geometrical perspective in the discrete problem of counting the number of solutions to an equation with inequality constraints. In fact, the geometry can be introduced before using it to visualize simple weighted-voting games. In two proofs without words (Haines and Jones [7],[9]), we relate the number of nonnegative integer solutions to $x + y + z = n$ with inequality constraints to triangular numbers. The purely visual approach in [7] mimics the Inclusion-Exclusion Principle, while our approach in [9] is not as direct. By viewing inequality constraints as separating the plane into half-planes, students explicitly see what solutions are being counted and eliminated in the inclusion-exclusion process. They then can make the connections from combinatorics to Euclidean space. This is a good lead in that helps students understand how the inequality constraints partition the simplex into different regions. Indeed, for three voters, the four possible Shapley-Shubik power indices and their permutations should be seen to exhaust the simplex. Exercises 23, 24, and 25 can be used to reinforce the concepts. We suggest using at least two class periods to introduce the geometry of simplices and the discretized simplex (where points are viewed as nonnegative integer solutions) and to apply the Inclusion-Exclusion Principle to physically partition the simplex into the equivalence classes.

Determining the likelihood that a randomly selected simple weighted-voting game with a fixed total weight has a particular Shapley-Shubik power index is a minor extension of the previous ideas. However, it is a little more challenging to show that this likelihood approaches the ratio of the area of the partition region to the area of the simplex, as was demonstrated in Example 6. Our solution to Example 6 simplifies the calculation by reconfiguring the problem geometrically. A more direct Inclusion-Exclusion Principle proof is also possible and may be more assessable and understandable for undergraduates. Students should have the intuition that as the total weight increases, then the number of dots in the simplex increases and fills up the region.

For a modeling course, probabilistic questions can be related to a player being a dictator, as in Proposition 4 and Corollary 5. These can easily be extended for any number of voters, *e.g.*, see [8]. This provides a good opportunity to discuss why a continuous approach is helpful when analyzing a discrete problem. As the counting becomes more complicated, the area can still be determined by geometry or integration. Hence, as long as the total weight is large enough, we expect the outcome of the Inclusion-Exclusion Principle (to determine the number of simple weighted-voting games in an equivalence class) to be proportional to the partition area on the simplex. Computing the probability using area may not be in the spirit of a combinatorics course, but it does show how to apply a variety of mathematical techniques to attack the same problem. And we believe that students need to see that there is more than one way to answer a question. Before unleashing your students on a problem to show the relationship between area and the total weight approaching infinity, it would be beneficial to explicitly solve a limit problem in class. Connecting areas and limits could take a good portion of two class periods.

We have assigned problems comparable to Example 6 as the kernel for a presentation by a masters level student, as part of a graduate combinatorial mathematics course. Another student presented geometrical approaches to 4-player simple weighted-voting games, where he viewed layers of the tetrahedron/3-simplex as triangles/2-simplices. He related his results for specific Shapley-Shubik power indices to triangular numbers as in [9], but could have used tetrahedral numbers, too. These students were comfortable thinking geometrically because visualization was emphasized throughout the course.

Just as writing and verbally expressing ideas are not introduced in a course at the expense of mathematical content, we believe that these examples/topics in voting theory can be integrated into the content without reducing the number of topics. In fact, just as communication enhances the understanding of mathematical content, we believe the recurrent use of simple weighted-voting games and the Shapley-Shubik power index also enhances the understanding.

3.4.7 Conclusion

As educators, we would like our students to be able to see how the material of a course can be applied to seemingly unrelated problems and that the tools developed in a mathematics course and the courses before it can be applied in many situations, not just those examples presented in textbooks. Sometimes texts revisit a particular application, but only using the mathematics from the one course. This paper presents a problem in voting theory that can be revisited throughout an undergraduate combinatorics class tying together ideas from combinatorics, geometry, calculus, and probability. Such a multi-perspective analysis leads to a deeper understanding of the mathematical thinking involved.

References

1. S. Berg, "On voting power indices and a class of probability distributions: With applications to EU data," *Group Decision and Negotiation*, 8 (1999) 17–31.
2. S.J. Brams and P.J. Affuso, "New Paradoxes of Voting Power on the EC Council of Ministers," *Electoral Studies*, 4 (1985) 135–139.
3. R. Brualdi, *Introductory Combinatorics*, 3rd ed., Prentice Hall, Upper Saddle River, NJ, 1999.

4. COMAP [Consortium for Mathematics and Its Applications], *For All Practical Purposes: Mathematical Literary in Today's World*, 5th ed., W.H. Freeman, New York, 2000.
5. J.S. Dreyer and A. Schotter, "Power relationships in the International Monetary Fund: The consequences of quota changes," *Review of Economics and Statistics*, 62 (1980) 97–106.
6. Euclid, *Thirteen Books of Euclid's Elements*, Heath, T. (translator and ed.), Dover, Mineola, NY, 1956.
7. M.J. Haines and M.A. Jones, "Four Proofs/No Words: Using Triangular Numbers to Count the Number of Nonnegative Integer Solutions to an Equality with Different Inequality Constraints," Preprint, 2002.
8. ——, "Geometric Implications of Power Indices and Apportionment Methods," Preprint, 2002.
9. ——, "Proof without Words: Nonnegative Integer Solutions and Triangular Numbers," *Mathematics Magazine*, V75, n5 (2002) 388.
10. M.J. Holler and G. Owen, editors, *Power Indices and Coalition Formation,* Kluwer Academic Publishers, Boston, 2001.
11. M.O. Hosli, "Admission of European Free Trade Association States to the European Community: Effect on voting power in the European Council of Ministers," *International Organization*, 47 (1993) 629–643.
12. J.P. Lampert, "Voting games, power indices, and presidential elections," *UMAP Module* (1988) 144–197.
13. I. Mann and L.S. Shapley, "The *a priori* voting strength of the Electoral College," in *Game Theory and Related Approaches to Social Behavior*, Shubik, M. (ed.), Wiley, New York, 1964.
14. H. Nurmi and T. Meskanen, "*A priori* power measures and the institutions of the European Union," *European Journal of Political Research,* 35 (1999) 161–179.
15. D.G. Saari, *Geometry of Voting*, Springer-Verläg, NY., 1994.
16. ——, *Basic Geometry of Voting*, Springer-Verläg, NY, 1995.
17. L.S. Shapley and M. Shubik, "A method for evaluating the distribution of power in a committee system," in *Game Theory and Related Approaches to Social Behavior*, Shubik, M. (ed.), Wiley, London, 1954.
18. P.D. Straffin, "Power indices in politics," in *Political and Related Models. Modules in Applied Mathematics*, V2, Springer-Verläg, New York, 1983.
19. ——, "Using Integrals to Evaluate Voting Power," *College Math Journal,* 10 (1979) 179–181.

Brief Biographical Sketches

Matthew J. Haines received a doctorate in arithmetic number theory from Lehigh University in 1994. More recently he has focused on math history, voting theory, and the preparation of elementary and secondary mathematics teachers. The NCTM standards emphasize that students should "recognize and use connections among mathematical ideas." Bringing history into the classroom helps students see the development of the interconnections, while voting theory is a mathematically rich topic with which the students can explore the use of these connections.

Michael A. Jones received his doctorate in game theory from the Mathematics Department at Northwestern University in 1994. He is a firm believer that research and teaching can feed off of one another: research makes teaching more effective and interesting for students while teaching focuses ones attention on a subject which acts as a filter for research ideas. His research interests are in the mathematics of the social sciences, including political science, economics, and psychology. He is particularly interested in analyzing discrete problems with continuous mathematics, as this article suggests.

3.5

An Innovative Approach to Post-Calculus Classical Applied Math

Robert J. Lopez
Rose-Hulman Institute of Technology (retired)
Maplesoft

3.5.1 Introduction

Post-calculus classical applied math is scattered through courses in differential equations, boundary value problems, vector calculus, matrix algebra, complex variables, and numerical methods. Most of this material can be found in texts entitled *Advanced Engineering Mathematics*. The mathematics in such texts is truly classical, having been available in its present format for many years, if not centuries. The apprenticeship for working in the field of classical applied mathematics is long and arduous because the apprentice must master material from so many different disciplines.

Twenty-first century software allows this apprenticeship to be both shorter and more effective. Modern computer algebra systems can be the tool of first-recourse for teaching, learning, and doing such applicable mathematics. Software tools such as Maple, Mathematica, MuPAD, and Macsyma implement nearly all the manipulations of the undergraduate program in applied and engineering mathematics. The time has come to use these twenty-first century tools for teaching eighteenth and nineteenth century mathematics.

A complete post-calculus applied math curriculum in which a computer algebra system is the primary working tool appears in [1]. In this text the software is not just an add-on to a traditional by-hands pedagogy. Instead, the software is used as an active partner in the student's participation in applied mathematics.

We give two examples taken from [1], examples that show how use of a computer algebra system enhances pedagogy. The purpose is not to tout a particular book, but instead, to call attention to the concept that a computer algebra system can, and should, be the working tool for teaching, learning, and doing classical applied math. So, rather than talk about this approach, we give two examples and let readers judge for themselves the viability of a curriculum predicated on the ubiquitous use of modern software tools.

3.5.2 Background Details

An ILI grant from NSF in 1988 brought computer algebra into the classroom at Rose-Hulman Institute of Technology (RHIT). By 1991 all calculus and differential equations courses were being taught in classrooms equipped with one computer per student. In 1995 RHIT implemented a laptop program that put a laptop computer into the hands of every new student. The introduction of computing hardware into the classroom was largely driven by the need to put computer algebra software into the hands of the students not only during class, but also during exams.

The students at RHIT are fairly homogeneous, since the school primarily provides undergraduate engineering and science programs. The first math course is calculus, although in more recent years a greater percentage of incoming students have had some high school exposure to calculus. At least two sections of 25 students enter with enough calculus to move directly into differential equations. However, as selective as RHIT is, its quarter system dictates a fairly lively pace, and students at every level find its programs demanding.

All students entering RHIT must have, or acquire, three quarters of calculus, up through and including multivariable calculus. However, vector calculus, including discussions of divergence and curl, and the integral theorems of Green, Stokes, and Gauss, is not covered. Engineering and most science students are required to take two quarters of differential equations, which, in recent years have included about half a quarter of matrix algebra in place of an equivalent exposure to boundary value problems. Computer science students are not required to take the second differential equations course, although many do as they often pursue a double major in both computer science and math.

With the advent of the first computer lab in 1988, the calculus and differential equations courses were revised to make use of computer algebra as a working tool. Almost immediately it was seen that formerly difficult topics such as finding eigenpairs for solving linear systems of ODEs, computing Fourier series, and solving boundary value problems became much easier for students. In 1992 the author was challenged by a publisher to write an advanced engineering math book based on the newly available computer tools. During the next five years as he debated undertaking this task, he developed courses in vector calculus, complex variables, numerical analysis, boundary value problems, and the calculus of variations, all using computer algebra as the primary working tool. As these courses were taught, the new approach of using symbolic computation as an active partner in pedagogy was being worked out.

An essential part of this development was the availability of computers during class, for homework, and during exams. Without this ubiquitous presence of the computer, the courses would have remained essentially by-hands, with a bit of computation grafted on as additional work. Having the computer available on exams meant that skills mastered in the computer environment would continue to be emphasized during exams, as well as in assignments. Creating a new pedagogy required complete access to a new set of working tools.

At the start of academic year 1997, a contract was signed for [1], and in the next two years a manuscript was written to capture the experiences of teaching post-calculus applied math course with a computer algebra system. Thus, RHIT's two courses in differential equations, its boundary value course, its linear algebra course, its two courses in numerical analysis, its course in vector calculus, and its course in complex variables all were captured in the manuscript. The only topics not included from an undergraduate engineering and science program are statistics, discrete signal processing, and the abstract algebra that appears in courses in discrete and combinatorial mathematics.

The drift of these remarks is to convince the reader that the new pedagogy is all inclusive. Computer algebra systems have evolved to the point where nearly all the mathematics an undergraduate must master can be implemented in such a system. But merely capturing the computations is not the significant issue. What we are convinced is really true is that learning mathematics with a computer algebra system is a richer, more efficient, and more effective learning experience.

To convince the reader of this same proposition, we will present two examples. The first will show how to use the convolution theorem for Laplace transforms to compute a convolution. Application of the definition of a convolution product can lead to delicate integrations that are rife with nested conditionals.

The convolution theorem allows the integration to be by-passed so that the convolution can be quickly evaluated. However, there are other tools available for analyzing the nested conditionals faced when evaluating the integrals generated by the definition of the convolution. We believe this interplay of approaches makes the convolution easier to comprehend.

The second example will show how to uncouple a first-order, constant-coefficient system of ordinary differential equations. The process of uncoupling the equations leads naturally to the diagonalization algorithm for the system matrix. This kind of investigative development is just too tedious to implement by hand, and is not experienced by the majority of students seeing the material for the first, and perhaps, only time. We believe this example shows how much richer the learning experience can be when illuminated with the right set of tools.

Before presenting these examples of how a computer algebra system changes pedagogy, we provide the following anecdotal evidence of improved learning under its aegis. When the prevailing technology was pencil and paper, Fourier series were not well received in the author's DE classes. Students would write some integral signs, generally not evaluate the integrals correctly, and conclude with a summation whose meaning was nonexistent. The topic was just a miasma of meaningless symbol-pushing. With the advent of the new technology students were able to evaluate integrals correctly, and could even graph partial sums of the resulting Fourier series to see if they converged to anything resembling the original function. The author's biggest surprise came when students brought assignments to his attention, asking why their partial sums did not seem to be converging to the appropriate function. When they made a computational error, they were in a position to realize it, and to seek the source of the error. It was on the basis of such experiences that the author maintained his enthusiasm for revising courses and pioneering a new pedagogy. Student reaction convinced the author that a more conceptual learning was taking place as a result of the use of modern computer technology in the classroom.

3.5.3 Example 1

As part of their mathematics requirements, students of engineering and applied mathematics typically meet the Laplace transform in a differential equations course. The course will generally include the convolution integrals and theorem. These students meet convolutions again in their engineering courses, either in electrical engineering or in control theory courses, where an intuitive approach is taken. The following example shows how theory, practice, and intuition can be developed simultaneously by the use of the appropriate software tools.

For the Laplace transform, convolution is defined by either of the integrals in

$$(f * g)(t) = \int_0^t f(t-x)g(x)dx = \int_0^t f(x)g(t-x)dx.$$

For the functions

$$f(t) = e^{-t} H(t) \quad \text{and} \quad g(t) = 2\sin(t-1)$$

where $H(t)$, the Heaviside function, is 0 or 1 accordingly as t is negative or positive, the first integral becomes

$$\int_0^t 2e^{-(t-x)} H(t-x) \sin(x-1) dx = e^{-t}(\cos 1 + \sin 1) + \sin(t-1) - \cos(t-1).$$

For an intuitive approach to understanding the convolution, engineers graph $f(x)$, $f(-x)$, and $f(t-x)$, obtaining Figures 3.5.1, 3.5.2, and 3.5.3, respectively. Figure 3.5.2 shows the reflection of $f(x)$ across the vertical axis, and Figure 3.5.3 shows $f(1-x)$, the translation of the reflected graph seen in Figure 3.5.2.

Figure 3.5.4 shows a graph of the factors $f(2.5-x)$ and $g(x)$, the first as the thick curve, the second as the thin. The product of these factors gives the integrand for the convolution at $t = 2.5$, graphed in Figure 3.5.5. The definite integral of this function on the interval $[0, 2.5]$ gives the value of the convolution at $t = 2.5$, a value obtained numerically as 1.04018444. If this process is repeated so as to form a succession of points, and the points plotted, we get Figure 3.5.6, a graph of the convolution.

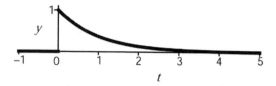

Figure 3.5.1. Graph of $f(t) = e^{-t} H(t)$

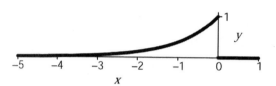

Figure 3.5.2. Graph of $f(-x)$ for the function f shown in Figure 3.5.1

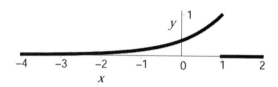

Figure 3.5.3. Graph of $f(1-x)$ for the function f shown in Figure 3.5.1

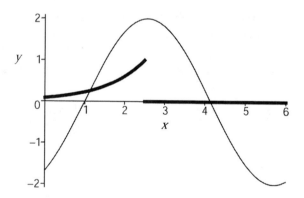

Figure 3.5.4. Graphs of $g(x) = \sin(x-1)$ (thin line), and of $f(2.5-x)$ (thick line) for f shown in Figure 3.5.1

As insightful as this process might be, it does not tell a student how to obtain a convolution. Instead, we suggest use of the convolution theorem for Laplace transforms whereby the Laplace transform of the convolution is the product of the transforms of the factors. Alternatively, we write

$$L^{-1}[F(s)G(s)] = f * g$$

3.5 An Innovative Approach to Post-Calculus Classical Applied Math

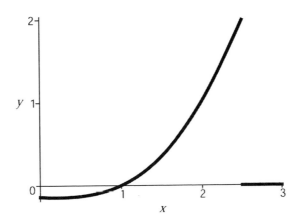

Figure 3.5.5. Graph of the product $f(2.5-x)*g(x)$, where f and g are shown in Figures 3.5.1 and 3.5.4, respectively.

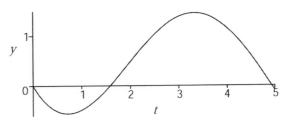

Figure 3.5.6. Graph of the convolution $(f*g)(t)$, where f and g are shown in Figures 3.5.1 and 3.5.4, respectively.

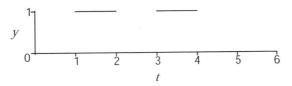

Figure 3.5.7. Graphs of the functions $f(t) = H(t-1) - H(t-2)$ and $g(t) = H(t-3) - H(t-4)$.

and compute

$$F(s)G(s) = \left[\frac{1}{s+1}\right]\left[2\frac{\cos 1 - s\sin 1}{s^2+1}\right] = \left[2\frac{\cos 1 - s\sin 1}{(s+1)(s^2+1)}\right]$$

$$f*g = L^{-1}[F(s)G(s)] = e^{-t}(\cos 1 + \sin 1) + \sin(t-1) - \cos(t-1)$$

With a computer algebra system, there is about as much work evaluating the convolution integral for these two functions as there is in using the convolution theorem. However, this is not always the case, as we see with the functions

$$f(t) = H(t-1) - H(t-2) \quad \text{and} \quad g(t) = H(t-3) - H(t-4)$$

whose graphs are seen in Figure 3.5.7. (The leftmost line segment is the nonvanishing portion of the graph of $f(t)$; the rightmost, $g(t)$.)

Evaluation of the Convolution by the Convolution Theorem

To compute the convolution $(f*g)(t)$ by Laplace transforms and the convolution theorem, obtain

$$L[f(t)] = \frac{e^{-s}-e^{-2s}}{s} \quad \text{and} \quad L[g(t)] = \frac{e^{-3s}-e^{-4s}}{s}$$

then invert the product of these transforms to obtain

$$L[f(t)]L[g(t)] = \frac{e^{-4s} - 2e^{-5s} + e^{-6s}}{s^2}$$

so that

$$(f * g)(t) = (t-4)H(t-4) - 2(t-5)H(t-5) + (t-6)H(t-6).$$

If we make the Heaviside function left-continuous by defining $H(0) = 0$, then the convolution can be written as

$$(f * g)(t) = \begin{cases} 0 & t \leq 4 \\ t - 4 & 4 < t \leq 5 \\ 6 - t & 5 < t \leq 6 \\ 0 & t > 6 \end{cases}$$

and its graph can be seen in Figure 3.5.8.

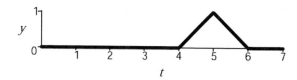

Figure 3.5.8. Graph of the convolution $(f * g)(t)$ for the functions shown in Figure 3.5.7.

Evaluation of the Convolution by the Convolution Integral

Evaluating the convolution integral

$$(f * g)(t) = \int_0^t [H(x-1) - H(x-2)][H(t-x-3) - H(t-x-4)]\,dx$$

requires knowing where in the xt-plane both $f(x)$ and $g(t-x)$ are simultaneously nonzero. Hence, the inequalities $x > 1$, $x < 2$, $t - x > 3$, and $t - x < 4$ must all be satisfied. The feasible region is shaded in Figure 3.5.9. The support of the integrand is the shaded parallelogram whose bounding edges are the lines $x = 1$, $x = 2$, $x = t - 3$, and $x = t - 4$, and whose vertices are the points $(1, 4)$, $(2, 5)$, $(2, 6)$, and $(1, 5)$. The shaded parallelogram is the region where the integrand has the value 1. Outside this region, the integrand has the value 0.

Figure 3.5.10 is an embellishment of Figure 3.5.9. Horizontal lines represent typical paths of integration along lines $t = $ constant. Any such line for which $t < 4$ or $t > 6$ yields a zero integrand and a value of zero for the convolution. Any such line between $t = 4$ and $t = 5$ yields an integrand of 1 and a value for the convolution that will be determined by the integral

$$\int_1^{t-3} 1\,dx = t - 4.$$

Any such line between $t = 5$ and $t = 6$ also yields an integrand of 1 and a value for the convolution that will be determined by the integral

$$\int_{t-4}^2 1\,dx = 6 - t.$$

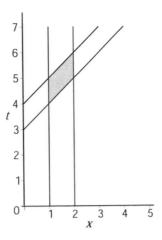

Figure 3.5.9. The region over which the integrand of the convolution integral is not zero for the functions in Figure 3.5.7.

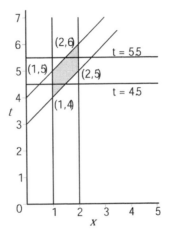

Figure 3.5.10. The region in Figure 3.5.9, along with representative horizontal lines of constant t.

The complete convolution can then be expressed as the piecewise function

$$(f * g)(t) = \begin{cases} 0 & t \leq 4 \\ t - 4 & 4 < t \leq 5 \\ 6 - t & 5 < t \leq 6 \\ 0 & t > 6 \end{cases}$$

in agreement with the results from the convolution theorem.

3.5.4 Example 2

Curricula for undergraduate science and engineering often contain units on both matrix algebra and systems of first order linear differential equations. The natural bridge from one to the other is the eigen-analysis of the coefficient matrix for the system. The following example shows how the gap between the theory and computation of eigenpairs and the solution of systems of differential equations can be closed in a natural and constructive way by the use of modern symbolic mathematics software.

Coupled Systems

The pair of differential equations

$$x'(t) = 18x(t) - 2y(t)$$
$$y'(t) = 12x(t) + 7y(t)$$

are *coupled* since both $x(t)$ and $y(t)$ appear in each equation. We apply the same term to the equivalent system $\mathbf{x}' = A\mathbf{x}$, where

$$A = \begin{bmatrix} 18 & -2 \\ 12 & 7 \end{bmatrix} \quad \text{and} \quad \mathbf{x} = \begin{bmatrix} x(t) \\ y(t) \end{bmatrix}. \tag{3.5.1}$$

It is this coupling which makes it difficult to solve the system.

Uncoupled Systems

The pair of differential equations

$$u'(t) = 10u(t) \quad \text{and} \quad v'(t) = 15v(t)$$

forms a system of the form $\mathbf{u}' = B\mathbf{u}$, where the diagonal matrix B and the vector $\mathbf{u}$ are respectively

$$B = \begin{bmatrix} 10 & 0 \\ 0 & 15 \end{bmatrix} \quad \text{and} \quad \mathbf{u} = \begin{bmatrix} u(t) \\ v(t) \end{bmatrix}.$$

These equations are said to be *uncoupled* since $u(t)$ appears only in the first equation while $v(t)$ appears only in the second. Uncoupled equations of this type are easily solved by separation of variables. Each of $u(t)$ and $v(t)$ are just constants times exponentials; specifically, they are

$$u(t) = c_1 e^{10t} \quad \text{and} \quad v(t) = c_2 e^{15t}.$$

Uncoupling Coupled Equations

Since uncoupled systems are so simple to solve, we ask if there is a way of uncoupling coupled systems. Consider the change of variables defined by

$$x(t) = a u(t) + b v(t)$$
$$y(t) = c u(t) + d v(t)$$

and expressed in matrix form by $\mathbf{x} = P\mathbf{u}$, where the matrix P is

$$P = \begin{bmatrix} a & b \\ c & d \end{bmatrix}.$$

Making this change of variables in the original coupled system $\mathbf{x} = A\mathbf{x}$, we obtain

$$a u'(t) + b v'(t) = (18a - 2c)u(t) + (18b - 2d)v(t)$$
$$c u'(t) + d v'(t) = (12a + 7c)u(t) + (12b + 7d)v(t)$$

which, in matrix notation is just $(P\mathbf{u})' = A(P\mathbf{u})$ or $\mathbf{u}' = (P^{-1}AP)\mathbf{u}$, where we isolated $\mathbf{u}'$ by multiplying through by P^{-1}, the inverse of P.

3.5 An Innovative Approach to Post-Calculus Classical Applied Math

We want our change of variables to result in a system of the form $\mathbf{u}' = B\mathbf{u}$, where B is a diagonal matrix. If A is the matrix on the left in (3.5.1), then we want

$$P^{-1}AP = \frac{1}{ad-bc}\begin{bmatrix} 18ad - 12ab - 2cd - 7bc & 11bd - 12b^2 - 2d^2 \\ 12a^2 - 11ac + 2c^2 & 12ab - 18bc + 2cd + 7ad \end{bmatrix}$$

to reduce to a diagonal matrix of the form

$$C = \begin{bmatrix} \alpha & 0 \\ 0 & \beta \end{bmatrix}$$

for some as-yet unknown values of α and β. Using an appropriate computer algebra system, the solutions of the four equations in the six unknowns a, b, c, d contained in $P^{-1}AP = C$, are found to be

$$a = \tfrac{c}{4} \quad b = \tfrac{2}{3}d \quad c = c \quad d = d \quad \alpha = 10 \quad \beta = 15$$
$$a = \tfrac{2}{3}c \quad b = \tfrac{d}{4} \quad c = c \quad d = d \quad \alpha = 15 \quad \beta = 10$$

There appear to be two possibilities, namely

$$C_1 = \begin{bmatrix} 10 & 0 \\ 0 & 15 \end{bmatrix} \quad \text{and} \quad P_1 = \begin{bmatrix} \tfrac{c}{4} & \tfrac{2}{3}d \\ c & d \end{bmatrix}$$

$$C_2 = \begin{bmatrix} 15 & 0 \\ 0 & 10 \end{bmatrix} \quad \text{and} \quad P_2 = \begin{bmatrix} \tfrac{2}{3}c & \tfrac{d}{4} \\ c & d \end{bmatrix}$$

For the diagonal matrices C_1 and C_2, the only difference is the order of the diagonal elements. It is no small surprise to find the numbers 10 and 15 along the diagonal! The corresponding solutions for the matrix P are really the same, except for the order of the columns. If, in P_1 we set $c = 4$ and $d = 3$, we get, except for column order, the same matrix as if we set $c = 3$ and $d = 4$ in matrix P_2.

The matrix P_1 diagonalizes the coupled system, resulting in the diagonal matrix C_1, while the matrix P_2 also diagonalizes the coupled system, resulting in the diagonal matrix C_2. The eigenvalues of A are 10 and 15, whereas the corresponding eigenvectors are

$$\begin{bmatrix} 1 \\ 4 \end{bmatrix} \quad \text{and} \quad \begin{bmatrix} 2 \\ 3 \end{bmatrix}$$

The columns of P_1 (or P_2) are multiples of the eigenvectors of A, and the diagonal entries of C_1 (or C_2) are the eigenvalues. Hence, knowledge of the eigenvalues and eigenvectors leads to an uncoupling, or diagonalization, of the system of differential equations. Alternatively, uncoupling the coupled system leads to the eigenpairs and the similarity transform by which the matrix A is diagonalized.

3.5.5 Conclusion

Using a computer algebra system, students can do more mathematics more efficiently. The premise of this paper is that teaching, learning, and doing applied mathematics via a computer algebra system is also more effective. The complete curriculum in undergraduate applied and engineering mathematics developed in [1] shows how twenty-first century computer software can be used to teach, learn, and do applied mathematics from the seventeenth through the twentieth centuries. The computer tools can (and are) used to examine relationships, amplifying what can be done with just a pencil and paper.

It is a mistake to graft modern software tools onto a traditional approach to calculus, differential equations, linear algebra, or any other of the courses that make up a program in classical applied mathematics. To teach this long-standing body of mathematics in the traditional by-hands fashion, merely paying lip-service to the new software tools, is an inadequate strategy. The new tools allow for a new apprenticeship and a new pedagogy. We really need to revise our texts to account for the impact of technology in the classroom. Ultimately, we want our texts to be as interactive and as engaging as a multimedia show. We cannot do this if we remain in the world of pencil and paper. But we can if we embrace modern software tools as the primary working tools for teaching, learning, and doing applied mathematics.

References

1. Robert J. Lopez, *Advanced Engineering Mathematics*, Addison-Wesley, Boston, 2001.

Brief Biographical Sketch

Robert J. Lopez is a classically trained applied mathematician with a Purdue University PhD (1970) in relativistic cosmology. After a short stint at the University of Nebraska-Lincoln, he spent twelve years at Memorial University in St. John's, Newfoundland, Canada. At the Rose-Hulman Institute of Technology, where he pioneered the use of *Maple* in the classroom, he authored books and papers, represented *Maple* on the road for 30 months, and received the Institute's awards for both teaching excellence and distinguished scholarship. He recently retired and returned to Canada to work full-time for Maplesoft.

About the Editor

Richard Maher received his PhD from Loyola University, Chicago. He served as department chair at Loyola from 1977 through 1985. His initial mathematical research involved the theory of lifting, and he has published a number of papers on this subject appearing in *Advances in Mathematics, Les Annales de L'Institut Fourier*, the *Journal of Mathematical Analysis and Applications, Rendiconti del Circolo Matematico Di Palermo*, and *Sudies in Probability and Ergodic Theory*. During his tour as chair, Professor Maher became interested in problems affecting undergraduate mathematical education, particularly those involving effective classroom methods. He has written a number of papers on this issue appearing in the *Notices (AMS), CBMS Issues in Mathematics Education, UME Trends*, and the MAA Notes series. He is also the author of the Calculus text, *Beginning Calculus with Applications*.

Teaching is the critical component of his professional activity, and for the past fifteen years, it has fueled his scholarly activity and interest.